Lecture Notes in Mathematics 1908

Editors:
J.-M. Morel, Cachan
F. Takens, Groningen
B. Teissier, Paris

Subseries:
École d'Été de Probabilités de Saint-Flour

T0255302

Saint-Flour Probability Summer School

The Saint-Flour volumes are reflections of the courses given at the Saint-Flour Probability Summer School. Founded in 1971, this school is organised every year by the Laboratoire de Mathématiques (CNRS and Université Blaise Pascal, Clermont-Ferrand, France). It is intended for PhD students, teachers and researchers who are interested in probability theory, statistics, and in their applications.

The duration of each school is 13 days (it was 17 days up to 2005), and up to 70 participants can attend it. The aim is to provide, in three high-level courses, a comprehensive study of some fields in probability theory or Statistics. The lecturers are chosen by an international scientific board. The participants themselves also have the opportunity to give short lectures about their research work.

Participants are lodged and work in the same building, a former seminary built in the 18th century in the city of Saint-Flour, at an altitude of 900 m. The pleasant surroundings facilitate scientific discussion and exchange.

The Saint-Flour Probability Summer School is supported by:

– Université Blaise Pascal
– Centre National de la Recherche Scientifique (C.N.R.S.)
– Ministère délégué à l'Enseignement supérieur et à la Recherche

For more information, see back pages of the book and
http://math.univ-bpclermont.fr/stflour/

Jean Picard
Summer School Chairman
Laboratoire de Mathématiques
Université Blaise Pascal
63177 Aubière Cedex
France

Terry J. Lyons
Michael Caruana · Thierry Lévy

Differential Equations Driven by Rough Paths

École d'Été de Probabilités
de Saint-Flour XXXIV - 2004

 Springer

Authors

Terry J. Lyons
Michael Caruana
Mathematical Institute
University of Oxford
24-29 St. Giles'
Oxford OX1 3LB
United Kingdom
e-mail: tlyons@maths.ox.ac.uk
 caruana@maths.ox.ac.uk

Thierry Lévy
Département de Mathématiques et
 Applications
École Normale Supérieure
45, rue d'Ulm
75230 Paris Cedex 05
France
e-mail: thierry.levy@ens.fr

Cover: Blaise Pascal (1623-1662)

Library of Congress Control Number: 2007922938

Mathematics Subject Classification (2000): 60H10, 60H05, 34A99, 60H30

ISSN print edition: 0075-8434
ISSN electronic edition: 1617-9692
ISSN Ecole d'Eté de Probabilités de St. Flour, print edition: 0721-5363
ISBN-10 3-540-71284-4 Springer Berlin Heidelberg New York
ISBN-13 978-3-540-71284-8 Springer Berlin Heidelberg New York
DOI 10.1007/978-3-540-71285-5

Springer is a part of Springer Science+Business Media
springer.com
© Springer-Verlag Berlin Heidelberg 2007

Typesetting by the authors and SPi using a Springer LaTeX macro package

Cover design: *design & production* GmbH, Heidelberg

Printed on acid-free paper   SPIN: 12031597   VA41/3100/SPi   5 4 3 2 1 0

A Word About the Summer School

The Saint-Flour Probability Summer School was founded in 1971. It is supported by CNRS, the "Ministère de la Recherche", and the "Université Blaise Pascal".

Three series of lectures were given at the 34th Probability Summer School in Saint-Flour (July 6–24, 2004), by the Professors Cerf, Lyons and Slade. We have decided to publish these courses separately. This volume contains the course of Professor Lyons; this final version has been written with two participants of the school, Michael Caruana and Thierry Lévy. We cordially thank them, as well as Professor Lyons for his performance at the summer school.

Sixty-nine participants have attended this school. Thirty-five of them have given a short lecture. The lists of participants and of short lectures are enclosed at the end of the volume.

Here are the references of Springer volumes which have been published prior to this one. All numbers refer to the *Lecture Notes in Mathematics* series, except S-50 which refers to volume 50 of the *Lecture Notes in Statistics* series.

1971: vol 307	1980: vol 929	1990: vol 1527	1998: vol 1738
1973: vol 390	1981: vol 976	1991: vol 1541	1999: vol 1781
1974: vol 480	1982: vol 1097	1992: vol 1581	2000: vol 1816
1975: vol 539	1983: vol 1117	1993: vol 1608	2001: vol 1837 & 1851
1976: vol 598	1984: vol 1180	1994: vol 1648	2002: vol 1840 & 1875
1977: vol 678	1985/86/87: vol 1362 & S-50	1995: vol 1690	2003: vol 1869 & 1896
1978: vol 774	1988: vol 1427	1996: vol 1665	2004: vol 1878, 1879 & 1908
1979: vol 876	1989: vol 1464	1997: vol 1717	2005: vol 1897

Further details can be found on the summer school web site
http://math.univ-bpclermont.fr/stflour/

Jean Picard
Clermont-Ferrand, September 2006

Foreword

I am ashamed to say that my first visit to the annual summer school at Saint Flour was to give the lectures recorded in the pages to follow. The attentive and supportive audience made the opportunity into a true privilege. I am grateful for the opportunity, to Jean Picard and his team for their considerable efforts and the audiences for their interest and patience.

I had been very busy in the weeks before the lectures and came equipped with a detailed outline and bibliography and a new tablet PC. Unlike my well-organised co-authors, who came with their printed notes, I wrote the detailed lecture notes as I went. Jean Picard, with his inevitable charm, tact and organisational skill, brought to the very rural setting of St Flour, a fast laser printer and every night there was a long period printing off and stapling the notes for the next days lecture (60 times over, as there was no Xerox machine!).

The notes were, in the main, hand-written on the tablet PC, supplemented by a substantial set of preprints and publications for the improvised library. This worked adequately and also provided some amusement for the audience: the computer placed a time stamp on each page permitting the audience to see which pages were written at 2.00 am and how long each page (that survived) took to write. Writing ten 90-min lectures in two weeks was a demanding but enjoyable task.

Two members of the meeting, Michael Caruana, and Thierry Lévy offered to convert the notes into the form you find here. They have taken my presentation to pieces, looked at it afresh, and produced a version that is cleaner and more coherent than I ever could have managed or imagined. I do not know how to express my gratitude. The original hand-written notes are, at the time of writing, to be found at:

http://sag.maths.ox.ac.uk/tlyons/st_flour/index.htm

The goal of these notes is to provide a straightforward and self-supporting but minimalist account of the key results forming the foundation of the theory of rough paths. The proofs are similar to those in the existing literature, but have been refined with the benefit of hindsight. We hope that the overall presentation optimises transparency and provides an accessible entry point into the theory without side distractions. The key result (the convergence of Picard Iteration and the universal limit theorem) has a proof that is significantly more transparent than in the original papers.

We hope they provide a brief and reasonably motivated account of rough paths that would equip one to study the published work in the area or one of the books that have or are about to appear on the topic.

Mathematical Goal

The theory of rough paths aims to create the appropriate mathematical framework for expressing the relationships between evolving systems, by extending classical calculus to the natural models for noisy evolving systems, which are often far from differentiable.

A rather simple idea works; differential equations obviously have meaning when used to couple smoothly evolving systems. If one could find metrics on smooth paths for which these couplings were uniformly continuous, then one could complete the space. The completions of the space of smooth paths are not complicated or too abstract and considering these spaces of "generalised paths" as the key spaces where evolving systems can be defined, modelled and studied seems fruitful. This approach has a number of applications, a few of which are mentioned in the notes. But the minimalistic approach we have set ourselves means we limit such discussion severely – the applications seem to still be developing and quite distinctive so we would commit the reader into much extra work and defeat the overall goal of this text. In Saint Flour it was natural to give probabilistic applications. The hand-written notes give the first presentation of a proof for a quite precise extension of the support theorem not reproduced here.

In 1936 Young introduced an extension to Stieltjes integration which applies to paths of p-variation less than 2. In a separate line of development, Chen (1957, geometry) and more recently Fliess (control theory), E. Platen (stochastic differential equations) and many others were lead to consider the sequence of iterated integrals of a path x in order to obtain a pathwise Taylor series of arbitrary order for the solution y to the vector equation

$$dy_t = f(y_t)\, dx_t.$$

These notes develop the non-commutative analysis required to integrate these two developments into the theory of rough paths, a mathematical framework for modelling the interaction between evolving systems.

Oxford, March 2006   *Terry J. Lyons*

Introduction

These notes put on record a series of ten lectures delivered by Terry Lyons at the Saint Flour summer school in July 2004. Terry Lyons's declared purpose was to bring the audience to the central result of the theory of rough paths along the straightest possible path. These notes, which follow closely the content of the lectures, are thus primarily intended for a reader who has never been exposed to the theory of rough paths. This introduction gives an overview of the subject and presents the content of each chapter, especially the first three, in some detail.

The theory of rough paths can be briefly described as a non-linear extension of the classical theory of controlled differential equations which is robust enough to allow a deterministic treatment of stochastic differential equations, or even controlled differential equations driven by much rougher signals than semi-martingales.

Let us first explain what a controlled differential equation is. In a setting where everything is differentiable, it is a differential equation of the form

$$\dot{Y}_t = F(\dot{X}_t, Y_t), \quad Y_0 = \xi, \tag{1}$$

where X is a given function, ξ is an initial condition, and Y is the unknown. The mapping F is taken to be linear with respect to its first variable. If F did not depend on its first variable, this would be the most general first-order time-homogeneous differential equation. The function F would be a vector field and Y would be the integral curve of this vector field starting at ξ. If now we had $\dot{Y}_t = F(t, Y_t)$ instead of (1), this would be the most general first-order time-inhomogeneous differential equation. The solution Y would be an integral curve of the time-dependent vector field F. Equation (1) is really of this kind, except that the time-inhomogeneity has been made explicitly dependent on a path X, which is said to control the equation. The physical meaning of (1) is the following: at each time, Y describes the state of a complex system (for instance the brain, or a car) and it evolves as a function of its present state and the infinitesimal variation of an external parameter (like the air pressure near the ears or the angle of the steering wheel).

Stochastic differential equations are of the form (1), except that X is usually far from being differentiable. They are often written under the form

$$dY_t = f(Y_t) \, dX_t, \quad Y_0 = \xi, \qquad (2)$$

which emphasises the linear dependence of the right-hand side with respect to dX_t. K. Itô, in introducing the concept of strong solution, emphasised the fact that the resolution of a stochastic differential equation amounts to the construction of a mapping between spaces of paths. For this reason, when the (deterministic) equation (2) admits a unique solution, we denote this solution by $Y = I_f(X, \xi)$, and we call I_f the Itô map associated with f.

The classical theory of differential equations tells us that if f is Lipschitz continuous, then (2) admits a unique solution as soon as X has bounded variation and this solution has bounded variation. Moreover, the Itô map I_f is continuous as a mapping between spaces of paths with bounded variation. The fundamental results of the theory of rough paths resolve the following two problems:

1. Identify a natural family of metrics on the space of paths with bounded variation such that the Itô map $X \mapsto I_f(X, \xi)$ is uniformly continuous with respect to these metrics, at least when f is regular enough.
2. Describe concretely the completion of the space of paths with bounded variation with respect to these metrics.

Let us give the solutions in a very condensed form. In the simplest setting, the appropriate metrics depend on a real parameter $p \in [1, +\infty)$ and two paths are close in the so-called *p-variation metric* if they are close in p-variation (a parameter-independent version of $\frac{1}{p}$-Hölder norm), as well as their first $\lfloor p \rfloor$ iterated integrals. An element of the completion of the space of paths with bounded variation on some interval $[0, T]$ with respect to the p-variation metric is called a *rough path* and consists in the data, for each sub-interval $[s, t]$ of $[0, T]$, of $\lfloor p \rfloor$ tensors – the first of which is the increment $x_t - x_s$ of some continuous path x – which summarise the behaviour of this rough path on $[s, t]$ in an efficient way, as far as controlled differential equations are concerned. This collection of tensors must satisfy some algebraic consistency relations and some analytic conditions similar to $\frac{1}{p}$-Hölder continuity.

The theorem which is proved in Chap. 5 of these notes states, under appropriate hypotheses, the existence and uniqueness of the solution of a differential equation controlled by a rough path.

It is worth noting that if the control X takes its values in a one-dimensional space, then the theory of rough paths becomes somehow trivial. Indeed, provided f is continuous, the Itô map is continuous with respect to the topology of uniform convergence and extends to the space of all continuous controls. The theory of rough paths is thus meaningful for *multi-dimensional* controls.

The coexistence of algebraic and analytic aspects in the definition of rough paths makes it somewhat difficult at first to get a general picture of the theory.

Chapters 1 and 2 of these notes present separately the main analytical and algebraic features of rough paths.

Chapter 1 is devoted to the concept of p-variation of a Banach-valued continuous function on an interval. Given a Banach space V and a real number $p \geq 1$, a continuous function $X : [0, T] \longrightarrow V$ is said to have finite p-variation if

$$\sup_{\mathcal{D}} \sum |X_{t_{i+1}} - X_{t_i}|^p < +\infty,$$

where the supremum is taken over all subdivisions $\mathcal{D}$ of $[0, T]$. This is equivalent to the fact that X can be reparametrised as a $\frac{1}{p}$-Hölder continuous function.

The central result of this chapter states that the classical Stieltjes integral $\int_0^t Y_u \, dX_u$, defined when Y is continuous and X has bounded variation, has an extension to the case where X and Y have finite p- and q-variation, respectively, provided $p^{-1} + q^{-1} > 1$. Moreover, in this case, the integral, as a function of t, has finite p-variation, like X. This was discovered by Young around 1930 and allows one to make sense of and even, if f is regular enough, to solve (2) when X has finite p-variation for $p < 2$.

On the other hand, if X is a real-valued path with finite p-variation for $p \geq 2$, then in general the Riemann sums $\sum X_{t_i}(X_{t_{i+1}} - X_{t_i})$ fail to converge as the mesh of the subdivision tends to 0 and one cannot anymore define $\int_0^t X_u \, dX_u$. One could think that this threshold at $p = 2$ is due to some weakness of the Young integral, but this is not the case. A simple and very concrete example shows that the mapping which to a path $X = (X_1, X_2)$: $[0, T] \longrightarrow \mathbb{R}^2$ with bounded variation associates the real number

$$\frac{1}{2} \int_0^T X_{1,u} \, dX_{2,u} - X_{2,u} \, dX_{1,u} \tag{3}$$

is *not* continuous in p-variation for $p > 2$.

The number defined by (3) is not just a funny counter-example. Firstly, it has a very natural geometric interpretation as the area enclosed by the curve X. Secondly, it is not very difficult to write it as the final value of the solution of a differential equation controlled by X, with a very regular, indeed a polynomial vector field f. So, even with a polynomial vector field, two paths which are close in p-variation for $p > 2$ do not determine close responses in the controlled equation (2). This is a first hint at the fact that it is natural to declare two paths with finite p-variation for $p > 2$ close to each other only if their difference has a small total p-variation and the areas that they determine are close.

It is interesting to note that stochastic differential equations stand just above this threshold $p = 2$. Indeed, almost surely Brownian paths have infinite two-variation and finite p-variation for every $p > 2$. The convergence in probability of the Riemann sums for stochastic integrals is, from the point of view of the Young integral, a miracle due to the very special stochastic structure of Brownian motion and the subtle cancellations it implies.

The quantity (3) has a sense as a stochastic integral when X is a Brownian motion: it is the Lévy area of X. It was conjectured by H. Föllmer that all SDEs driven by a Brownian motion can be solved at once, i.e. outside a single negligible set, once one has chosen a version of the Lévy area of this Brownian motion. The theory of rough paths gives a rigourous framework for this conjecture and proves it.

Chapter 2 explores the idea that (3) is the first of an infinite sequence of quantities which are canonically associated to a path with bounded variation in a Banach space. These quantities are the iterated integrals of the path. Let V be a Banach space. Let $X : [0,T] \longrightarrow V$ be a path with bounded variation. For every integer $n \geq 1$, and every (s,t) such that $0 \leq s \leq t \leq T$, the nth iterated integral of X over $[s,t]$ is the tensor of $V^{\otimes n}$ defined by

$$X_{s,t}^n = \int_{s<u_1<\ldots<u_n<t} dX_{u_1} \otimes \ldots \otimes dX_{u_n}. \tag{4}$$

When V is $\mathbb{R}^2$, (3) is just the antisymmetric part of $X_{0,T}^2$. The zeroth iterated integral is simply $X_{s,t}^0 = 1 \in \mathbb{R} = V^{\otimes 0}$.

The importance of iterated integrals has been recognised by geometers, in particular K.T. Chen, a long time ago. In the context of controlled differential equations, their importance is most strikingly illustrated by the case of linear equations. Linear equations are those of the form (2) where the vector field f depends linearly on Y. When the control X has bounded variation, the resolution of the equation by Picard iteration leads to an expression of the solution Y as the sum of an infinite series of the form

$$Y_t = \left(\sum_{n=0}^{\infty} f^n X_{0,t}^n \right) Y_0, \tag{5}$$

where f^n is an operator depending on f and n and whose norm grows at most geometrically with n. It is not hard to check that the norm of the iterated integral $X_{0,t}^n$ of X decays like $\frac{1}{n!}$. Thus, the series (5) converges extremely fast. In typical numerical applications, a dozen of terms of the series suffice to provide an excellent approximation of the solution. What is even better is the following: once a dozen of iterated integrals of X have been stored on a computer, i.e. around d^{12} numbers if d is the dimension of V, it is possible to solve numerically very accurately any linear differential equation controlled by X with very little extra computation. The numerical error can be bounded by a simple function of the norm of the vector field.

The iterated integrals of X over an interval $[s,t]$ are thus extremely efficient statistics of X, in the sense that they determine very accurately the response of any linear system driven by X. It is in fact possible to understand exactly what it means geometrically for two paths with bounded variation and with the same origin to have the same iterated integrals: it means that they differ by a tree-like path. We state this result precisely in the notes, but do not include its proof.

However satisfying the results above are: one is missing something essential about iterated integrals until one considers them all as a single object. This object is called the *signature of the path*. More precisely, with the notation used in (4), the signature of X over the interval $[s, t]$ is the infinite sequence in $\mathbb{R} \oplus V \oplus V^{\otimes 2} \oplus \ldots$ defined as follows:

$$S(X)_{s,t} = (1, X^1_{s,t}, X^2_{s,t}, \ldots). \tag{6}$$

The infinite sequence space $\oplus_{n \geq 0} V^{\otimes n}$ is called the *extended tensor algebra* of V and it is denoted by $T((V))$. It is indeed an algebra for the multiplication induced by the tensor product. The reader not familiar with this kind of algebraic structures should keep the following dictionary in mind. Assume that V has finite dimension d and choose a basis $(v_1, \ldots, v_d)$ of V. Then it is a tautology that $V = V^{\otimes 1}$ is isomorphic to the space of homogeneous polynomials of degree 1 in the variables $X_1, \ldots, X_d$. It turns out that for every $n \geq 0$, $V^{\otimes n}$ is isomorphic to the space of homogeneous polynomials of degree n in the *non-commuting* variables $X_1, \ldots, X_d$. For instance, if $d = 2$, then a basis of $V^{\otimes 2}$ is $(X_1^2, X_1 X_2, X_2 X_1, X_2^2)$. Finally, $T((V))$ is isomorphic to the space of all formal power series in d non-commuting variables, not only as a vector space, but also as an algebra: the product of tensors corresponds exactly to the product of non-commuting polynomials.

The fundamental property of the signature is the following: if (s, u, t) are such that $0 \leq s \leq u \leq t \leq T$, then

$$S(X)_{s,t} = S(X)_{s,u} \otimes S(X)_{u,t}. \tag{7}$$

This multiplicativity property, although it encodes infinitely many relations between iterated integrals of X, can be proved in a very elementary way. However, the following abstract and informal point of view gives an interesting insight on the signature of a path. Among all differential equations that X can control, there is one which is more important than the others and in a certain sense universal. It is the following:

$$dS_t = S_t \otimes dX_t, \quad S_0 = 1, \quad S : [0, T] \longrightarrow T((V)). \tag{8}$$

The solution to (8) is nothing but the signature of X: $S_t = S(X)_{0,t}$. This suggests that the signature of a path should be thought of as a kind of universal non-commutative exponential of this path. Moreover, we deduce from (8) that the two sides of (7), which satisfy the same differential equation with the same initial value, are equal.

Chapter 3 focuses on collections $(S_{s,t})_{0 \leq s \leq t \leq T}$ of elements of $T((V))$ which satisfy (7). Such collections are called *multiplicative functionals* and the point of the theory of rough paths is to take them as the fundamental objects driving differential equations. Rough paths are multiplicative functionals which satisfy some regularity property related to p-variation.

Like the multiplicativity property, the regularity property is inspired by the study of the signature of a path. If X has bounded variation, then $X^1_{s,t} = X_t - X_s \sim |t - s|$ and $X^n_{s,t}$ is of the order of $|t - s|^n$. If X has only finite p-variation for some $p \in (1, 2)$, then it is still possible to define its iterated integrals as Young integrals, and we expect $X^n_{s,t}$ to be of the order of $|t - s|^{\frac{n}{p}}$.

Let $\triangle_T$ denote the set of pairs (s, t) such that $0 \leq s \leq t \leq T$. A multiplicative functional of degree n in V is a continuous mapping $X : \triangle_T \longrightarrow T^{(n)}(V) = \bigoplus_{i=0}^{n} V^{\otimes i}$. For each $(s, t) \in \triangle_T$, $X_{s,t}$ is thus a collection of $n + 1$ tensors $(1, X^1_{s,t}, X^2_{s,t}, \ldots, X^n_{s,t})$. Let $p \geq 1$ be a number. A multiplicative functional X is said to have finite p-variation if

$$\sup_{0 \leq i \leq n} \sup_{\mathcal{D}} \sum |X^i_{t_k, t_{k+1}}|^{\frac{p}{i}} < +\infty. \tag{9}$$

The first fundamental result of the theory expresses a deep connection between the multiplicativity property (7) and the finiteness of the p-variation (9). It states that a multiplicative functional with finite p-variation is determined by its truncature at level $\lfloor p \rfloor$. More precisely, if X and Y are multiplicative functionals of degree $n \geq \lfloor p \rfloor$ with finite p-variation and $X^i_{s,t} = Y^i_{s,t}$ for all $(s, t) \in \triangle_T$ and $i = 0, \ldots, \lfloor p \rfloor$, then $X = Y$. Conversely, any multiplicative functional of degree m with finite p-variation can be extended to a multiplicative functional of arbitrarily high degree with finite p-variation, provided $\lfloor p \rfloor \leq m$.

A p-rough path is then defined to be a multiplicative functional of finite p-variation and degree $\lfloor p \rfloor$.

Chapters 4 and 5 give a meaning to differential equations driven by rough paths and present a proof of the main theorem of the theory, named Universal Limit Theorem by P. Malliavin, which asserts that, provided f is smooth enough, (2) admits a unique solution when X is a p-rough path. The solution Y is then itself a p-rough path.

Let us conclude this introduction by explaining the part of the Universal Limit Theorem which can be stated without referring to rough paths, i.e. let us describe the metrics on the space of paths with bounded variation with respect to which the Itô map is uniformly continuous. Choose $p \geq 1$. Consider X and $\tilde{X}$, both with bounded variation. For all (s, t) and all $n \geq 0$, let $X^n_{s,t}$ and $\tilde{X}^n_{s,t}$ denote their iterated integrals of order n over $[s, t]$. Then the distance in the p-variation metric between X and $\tilde{X}$ is defined by

$$d_p(X, \tilde{X}) = \sup_{0 \leq i \leq \lfloor p \rfloor} \sup_{\mathcal{D}} \left[\sum |X^i_{t_k, t_{k+1}} - \tilde{X}^i_{t_k, t_{k+1}}|^{\frac{p}{i}} \right]^{\frac{1}{p}}. \tag{10}$$

Now, provided the vector field f is of class $C^{\lfloor p \rfloor + \varepsilon}$, the Itô map I_f is uniformly continuous with respect to the distance d_p.

It has been a great pleasure for us to write these notes, not the least thanks to the countless hours of discussions we have had with Terry Lyons during their preparation. We would like to thank him warmly for his kind patience and communicative enthusiasm. Our greatest hope is that some of this enthusiasm, together with some of our own fascination for this theory, will permeate through these notes to the reader.

Contents

1

Differential Equations Driven by Moderately Irregular Signals

This chapter is not properly speaking about rough paths. It presents the first of the two threads that constitute the heart of the theory of rough paths, namely the Young integral. This integral, which is not in the scope of the classical measure theory, allows one to solve non-autonomous differential equations driven by certain paths with unbounded variation. This is a significant and powerful extension of the classical theory of ordinary differential equations. The main concept involved in this extension is that of p-variation of a function, which we discuss extensively. However, the approach to ordinary differential equations based on the Young integral has its limitations and it will appear that it is essentially unsuitable for controls which are as irregular as typical Brownian paths. Another idea is needed to handle such rough controls, and this will be the object of Chap. 2.

1.1 Signals with Bounded Variation

1.1.1 The General Setting of Controlled Differential Equations

Let V and W be two Banach spaces. Let $\mathbf{L}(V,W)$ denote the set of continuous linear mappings from V to W. Let $J = [0,T]$ be a compact interval. Let $X : J \longrightarrow V$, $Y : J \longrightarrow W$ and $f : W \longrightarrow \mathbf{L}(V,W)$ be continuous mappings.

Assume that X, Y and f are regular enough for the integral $\int_0^t f(Y_s)\, dX_s$ to make sense for each $t \in J$. For example, it is enough to assume that X is Lipschitz continuous, but we shall discuss a whole range of other possible assumptions. Let ξ be an element of W. We say that X, Y, f satisfy the equation

$$dY_t = f(Y_t)\, dX_t, \quad Y_0 = \xi, \tag{1.1}$$

if the following equality holds for all $t \in J$:

$$Y_t = \xi + \int_0^t f(Y_s)\, dX_s.$$

In this case, f is called the *vector field*, X the *signal*, or *control*, or *driving noise*, or *input*, ξ the *initial condition*, and Y the *solution*, or *response*, or *output*. This terminology alone accounts for the importance of (1.1).

Solving (1.1) amounts to finding a fixed point for the functional $Y_\bullet \mapsto \xi + \int_0^\bullet f(Y_s)\, dX_s$. This requires an understanding of the elementary functionals $Y \mapsto f(Y)$ and $Y \mapsto \int Y\, dX$ as nice mappings between appropriate spaces of functions. A hint at the nature of the relevant spaces of functions is the following property of invariance under reparametrisation: if $Y_\bullet$ is a solution to (1.1) driven by $X_\bullet$ and if $\varphi : J \longrightarrow J$ is a non-decreasing surjection, then $(Y \circ \varphi)_\bullet$ is a solution to the same equation driven by $(X \circ \varphi)_\bullet$. Hence, we expect properties of functions that are invariant under reparametrisation to play an important role.

Once the mapping $Y \mapsto \int f(Y)\, dX$ is understood, classical strategies for finding fixed points apply, in particular *iteration* which, when it works, usually also provides uniqueness of the solution and, to some extent, its regular dependence with respect to the control and the initial condition. Under assumptions too weak for iteration to apply, more sophisticated fixed point theorems can sometimes be used to prove the existence of a solution.

Remark 1.1. The definition of f as a continuous mapping from W to $\mathbf{L}(V, W)$ makes the notation $f(Y_s)\, dX_s$ transparent. However, we want to emphasise that there is another equivalent interpretation of f which, we believe, is more appropriate.

Let $C(W, W)$ denote the linear space of continuous mappings from W to itself. The elements of $C(W, W)$ are nothing but continuous vector fields on W. Then f can be thought of as an element of $\mathbf{L}(V, C(W, W))$. From this point of view, f is a linear form on V with values in the space of vector fields on W. In words, f transforms an infinitesimal displacement in the space of X's into a vector field on the space of Y's, i.e. an infinitesimal transformation of the whole space of Y's.

This picture may be expressed in a slightly more formal way as follows (the reader unfamiliar with differential geometry can safely skip the end of this remark). Assume that the signal X takes its values in some manifold M and the response in a manifold N. Let $\mathcal{F}(N)$ denote the linear space of vector fields on N. Then f is a one-form on M with values in $\mathcal{F}(N)$. Thinking informally of $\mathcal{F}(N)$ as the Lie algebra of the group $\mathrm{Diff}(N)$ of diffeomorphisms of N, we can see (1.1) as defining the development of X in $\mathrm{Diff}(N)$ starting at the identity.

1.1.2 The Theorems of Picard–Lindelöf and Cauchy–Peano

Let us illustrate this general setting in a classical case.

Definition 1.2. *Let $J = [0, T]$ be a closed bounded interval. A partition or a subdivision of J is an increasing sequence of real numbers $\mathcal{D} = (t_0, t_1, \ldots, t_r)$*

such that $0 \le t_0 < t_1 < \ldots < t_r \le T$. We use the notation $\mathcal{D} \subset J$ to indicate that $\mathcal{D}$ is a partition of J.

Recall that the path $X : J \longrightarrow V$ is said to have *bounded variation* if

$$\sup_{\mathcal{D} \subset J} \sum_{i=0}^{r-1} |X_{t_{i+1}} - X_{t_i}| < +\infty.$$

Let us assume that X has bounded variation. Since f and Y are continuous, the integral $\int_0^t f(Y_s) \, dX_s$ exists for all $t \in J$ as a Stieltjes integral.

Theorem 1.3 (Picard–Lindelöf). *Assume that X has bounded variation and f is Lipschitz continuous. Then, for every $\xi \in W$, the differential equation (1.1) admits a unique solution.*

The main argument for proving this theorem is that, if $[s,t]$ is a small enough sub-interval of J, then the mapping $Y_\bullet \mapsto \xi + \int_s^\bullet f(Y_u) \, dX_u$ from $C^0([s,t], W)$ to itself is a contraction in the uniform norm. Local existence and uniqueness follow by the classical fixed point theorem. Global existence and uniqueness can then be proved by subdividing J into finitely many small sub-intervals on which the local argument applies.

Let $I_f(X, \xi)$ denote the unique solution to (1.1), where I stands for *Itô map*. Then a closer look at the iteration procedure which leads to a solution allows one to prove that $I_f(X, \xi)$ depends continuously on X and ξ. More precisely, the total variation of $I_f(X', \xi') - I_f(X, \xi)$ can be made arbitrarily small by choosing on one hand ξ' close enough to ξ in W, and on the other hand X' such that the total variation of $X' - X$ on J is small enough.

If we drop the assumption that f is Lipschitz continuous, then we lose the uniqueness of the solution.

Theorem 1.4 (Cauchy–Peano). *Assume that W is finite-dimensional. Assume that X has bounded variation and f is continuous. Then, for every $\xi \in W$, the differential equation (1.1) admits a solution.*

This result relies on Schauder's fixed point theorem, which goes as follows. Let E be a Banach space and $B \subset E$ a convex subset. Let $F : E \longrightarrow E$ be a continuous mapping such that $F(B) \subset B$ and $F(B)$ is a relatively compact subset of E. Then F admits a fixed point in B.

In the context of Cauchy–Peano's theorem, take B to be a large ball in uniform norm in the space E of W-valued continuous functions on a small sub-interval of J. Let $F : E \longrightarrow E$ be defined by $F(Y)_\bullet = \int^\bullet f(Y) \, dX$. If the ball is large enough and the sub-interval small enough, then $F(B) \subset B$. On the other hand, the subset $F(B)$ of E is bounded and uniformly equicontinuous, hence relatively compact by Ascoli's theorem. The local existence of a solution follows and the global existence is proved once again by subdividing J into finitely many sub-intervals on which the local argument applies.

After constructing the Young integral, we will be able to prove two theorems very similar to Picard's and Peano's theorem in the context of paths with finite p-variation, namely Theorems 1.20 and 1.28. Their proofs follow closely the strategies we have just explained, and we will give them with much more details.

1.2 Paths with Finite p-Variation

1.2.1 Definitions

Let us choose once for all in this section a Banach space E, whose norm is denoted by $|\cdot|$. Let us also fix a compact interval J. Following N. Wiener for the case $p = 2$ and Young [9] for the general case, let us define the p-variation of a E-valued path on J.

Definition 1.5. *Let $p \geq 1$ be a real number. Let $X : J \longrightarrow E$ be a continuous path. The p-variation of X on the interval J is defined by*

$$\|X\|_{p,J} = \left[\sup_{\mathcal{D} \subset J} \sum_{j=0}^{r-1} |X_{t_j} - X_{t_{j+1}}|^p \right]^{\frac{1}{p}}.$$

This definition is perhaps deceitfully simple. Let us emphasise that we consider the *supremum over all subdivisions* of J, not a limit as the mesh of the subdivision tends to zero. Unless $p = 1$, this is quite different. For example, if $p > 1$, then the reader should convince himself that, given any continuous path $X : J \longrightarrow \mathbb{R}$, not necessarily of bounded variation, there exists a sequence of subdivisions of J whose mesh tends to zero and along which the p-variation of X tends to 0 [34]. Nevertheless, $\|X\|_{p,J}$ can be equal to zero only if X is constant.

Let us give a few examples. A path with finite one-variation is just a path with bounded variation, also called a path with finite length. If $E = \mathbb{R}$, then a continuous path has finite 1-variation if and only if it is of the form $X_\bullet = \int_0^\bullet \mu(dt)$, where μ is a signed measure with no atoms and finite total mass.

A path which is Hölder continuous with exponent α, with $0 < \alpha \leq 1$, has finite $\frac{1}{\alpha}$-variation. Hence, a typical Brownian path has finite p-variation for every $p > 2$. However, the 2-variation of a Brownian path on $[0, T]$ is almost surely infinite [34, p. 48]. On the other hand, if $(B_t)_{t\in[0,T]}$ is a Brownian motion, then the path $t \mapsto B_t$ in $L^2(\Omega, P)$ has finite two-variation on $[0, T]$, equal to T.

The following properties of the p-variation are elementary.

Lemma 1.6. *Let $X : J \longrightarrow E$ be a continuous path.*

1. *Let $\varphi : J \longrightarrow J$ be a non-decreasing surjection. Then, for all $p \geq 1$, $\|X\|_{p,J} = \|X \circ \varphi\|_{p,J}$.*
2. *The function $p \mapsto \|X\|_{p,J}$ from $[1, +\infty)$ to $[0, +\infty]$ is non-increasing.*

3. The function $p \mapsto \log \|X\|_{p,J}^p$ is convex, and continuous on any interval where it is finite.
4. For all $p \geq 1$, $\|X\|_{p,J} \geq \sup_{s,t \in J} |X_s - X_t|$.

For each $p \geq 1$, let $\mathcal{V}^p(J, E)$ or simply $\mathcal{V}^p$ denote the subset of $C^0(J, E)$ consisting of those paths which have finite p-variation. For each $X \in \mathcal{V}^p(J, E)$, set

$$\|X\|_{\mathcal{V}^p(J,E)} = \|X\|_{p,J} + \sup_{t \in J} |X_t|.$$

To avoid any ambiguity, we call $\|X\|_{p,J}$ the p-variation of X on J and $\|X\|_{\mathcal{V}^p(J,E)}$ the p-variation norm of X on J.

Proposition 1.7. For each $p \geq 1$, the set $\mathcal{V}^p(J, E)$ is a linear subspace of $C^0(J, E)$ on which $\| \cdot \|_{\mathcal{V}^p(J,E)}$ is a norm. Moreover, $(\mathcal{V}^p(J, E), \| \cdot \|_{\mathcal{V}^p(J,E)})$ is a Banach space. Finally, if $1 \leq p \leq q$, then the inclusions

$$\mathcal{V}^1(J, E) \subset \mathcal{V}^p(J, E) \subset \mathcal{V}^q(J, E) \subset C^0(J, E)$$

hold and are continuous.

Proof. Let X and Y be two elements of $\mathcal{V}^p(J, E)$. It follows immediately from the definition of the p-variation that $\|X + Y\|_{p,J} \leq \|X\|_{p,J} + \|Y\|_{p,J}$. This implies that $\mathcal{V}^p(J, E)$ is indeed a linear subspace of $C^0(J, E)$ and that $\| \cdot \|_{\mathcal{V}^p(J,E)}$ is a norm on $\mathcal{V}^p(J, E)$. The rest of the proof is standard. $\square$

We prove now that the p-variation of continuous paths is lower semi-continuous with respect to the topology of pointwise convergence.

Proposition 1.8 (Lower Semi-Continuity). Let $(X(n))_{n \geq 0}$ be a sequence of elements of $C^0(J, E)$ which converges pointwise to a continuous path X. Then

$$\liminf_{n \to \infty} \|X(n)\|_{p,J} \geq \|X\|_{p,J}.$$

Proof. Choose $\varepsilon > 0$. Let $\mathcal{D}_\varepsilon$ be a partition of J such that

$$\left[\sum_{\mathcal{D}_\varepsilon} |X_{t_j} - X_{t_{j+1}}|^p \right]^{\frac{1}{p}} \geq \|X\|_{p,J} - \varepsilon.$$

Since $\mathcal{D}_\varepsilon$ is finite, we have

$$\lim_{n \to \infty} \left[\sum_{\mathcal{D}_\varepsilon} |X_{t_j}(n) - X_{t_{j+1}}(n)|^p \right]^{\frac{1}{p}} \geq \|X\|_{p,J} - \varepsilon.$$

The result follows by letting ε tend to 0. $\square$

According to this result, for all $p \geq 1$, the set $\{X \in \mathcal{V}^p(J,E) \mid \|X\|_{p,J} \leq 1\}$ is closed in the uniform topology. It is easy to check that it is not compact, but we shall prove in the next paragraph that from any sequence with uniformly bounded p-variation, one can extract a subsequence which converges uniformly *after reparametrisation*.

The stability of the space $\mathcal{V}^p(J,E)$ by reparametrisation has a very important consequence: unless $E = \{0\}$ or J is a single point, it is never separable. For example, when $p = 1$, consider two mutually singular probability measures without atoms μ and ν on $J = [0,1]$ and set $X_\bullet = \int_0^\bullet d\mu$ and $Y_\bullet = \int_0^\bullet d\nu$. Then X and Y belong to $\mathcal{V}^1(J,\mathbb{R})$ and $\|X - Y\|_{1,[0,1]} = 2$. The fact that there exists uncountably many mutually singular probability measures without atoms on $[0,1]$ can be proved by considering the normalised Hausdorff measures on an uncountable family of subsets of $[0,1]$ with pairwise distinct Hausdorff dimensions. For instance, for every $\xi \in (0,\frac{1}{2})$, the perfect symmetric set C_ξ with constant ratio ξ is built by iterating the transformation $[0,1] \mapsto [0,\xi] \cup [1-\xi,1]$. The Hausdorff dimension of C_ξ is $-\log(2)/\log(\xi)$ and in this dimension, the Hausdorff measure of C_ξ is positive and finite.

1.2.2 Controls

Consider a path X in $\mathcal{V}^p(J,E)$ for some $p \geq 1$. For all $s \leq t$ belonging to J, define
$$\omega_X(s,t) = \|X\|_{p,[s,t]}^p = \sup_{\mathcal{D} \subset [s,t]} \sum_{i=0}^{r-1} |X_{t_i} - X_{t_{i+1}}|^p.$$

The function $(s,t) \mapsto \omega_X(s,t)$ is non-negative and vanishes on the diagonal $s = t$. It is non-decreasing in t and non-increasing in s. It is in fact *super-additive*, in the sense that, for all $s \leq t \leq u$ in J,
$$\omega_X(s,t) + \omega_X(t,u) \leq \omega_X(s,u).$$

Finally, it is continuous in s and t. All these properties are easy to prove, but the fact that, as $s_n \uparrow s$ and $t_n \downarrow t$, $\omega_X(s_n,t) \to \omega_X(s,t)$ and $\omega_X(s,t_n) \to \omega_X(s,t)$ requires some care.

The function $\omega_X(0,\cdot)$ provides us with a natural reparametrisation for X. Indeed, assuming that X is constant on no sub-interval of J, the function $t \mapsto \omega_X(0,t)\frac{T}{\omega_X(0,T)}$ is an increasing bijection $J \longrightarrow J$. Let $t \mapsto \tau(t)$ be its inverse. Then, for all $s \leq t$ in J,
$$|X_{\tau(s)} - X_{\tau(t)}|^p \leq \omega_X(\tau(s),\tau(t)) \leq \omega_X(0,\tau(t)) - \omega_X(0,\tau(s)) = \frac{\omega_X(0,T)}{T}(t-s).$$

We have just proved that any path with finite p-variation can be reparametrised as a Hölder continuous path with exponent $\frac{1}{p}$. In other words, paths with finite p-variation are Hölder continuous paths with exponent $\frac{1}{p}$ *up to reparametrisation*. Moreover, the Hölder norm can be chosen uniformly on sets of paths with uniformly bounded p-variation. In particular, such sets can be reparametrised so as to become uniformly equicontinuous.

We now define an abstract control function. For this, let $\triangle_T$ denote the simplex $\{(s,t) \in [0,T]^2 : 0 \leq s \leq t \leq T\}$.

Definition 1.9. *A control function, or control, on $J = [0,T]$ is a continuous non-negative function ω on $\triangle_T$ which is super-additive in the sense that*

$$\omega(s,t) + \omega(t,u) \leq \omega(s,u) \quad \forall s \leq t \leq u \in J$$

and for which $\omega(t,t) = 0$ for all $t \in J$.

It is plain that the function ω_X defined above is a control function. The sum of finitely many controls is still a control. If the sum of infinitely many controls converges pointwise on $\triangle_T$, then this sum is still a control. However, the maximum of two controls is not necessarily a control because it may fail to be super-additive.

Lemma 1.10 is a straightforward consequence of the super-additivity.

Lemma 1.10. *Let ω be a control. Let $X : J \longrightarrow E$ be a continuous path. Assume that, for some $p \geq 1$ and for all $(s,t) \in \triangle_T$, one has $|X_s - X_t|^p \leq \omega(s,t)$. Then, for all $(s,t) \in \triangle_T$, $\|X\|_{p,[s,t]} \leq \omega(s,t)^{\frac{1}{p}}$.*

When the conclusion of this lemma holds, we say that the p-variation of X is *controlled by* ω.

1.2.3 Approximation Results

Let us investigate now how paths with finite p-variation can be approximated by piecewise linear paths.

Consider first the special case of real-valued paths with finite length. Any such path can be reparametrised in such a way that it becomes Lipschitz continuous, in particular absolutely continuous. We say that a path $X \in \mathcal{V}^1(J, \mathbb{R})$ has absolutely continuous one-variation if the function $t \mapsto \|X\|_{1,[0,t]}$ is absolutely continuous. This is equivalent to saying that it is the primitive of a signed measure which is absolutely continuous with respect to Lebesgue measure.

Proposition 1.11. *Let $(\mathcal{D}_n)_{n \geq 0}$ be an increasing sequence of subdivisions of J such that the mesh of $\mathcal{D}_n$ tends to 0 as n tends to ∞. Let X be a real-valued path on J with absolutely continuous one-variation. For each $n \geq 0$, let $X(n)$ be the piecewise linear approximation of X corresponding to $\mathcal{D}_n$. Then the sequence $(X(n))_{n \geq 0}$ converges to X in one-variation.*

Since $\mathcal{V}^1(J, \mathbb{R})$ is not separable, we cannot expect such a result to hold if we drop the assumption that X has *absolutely continuous* one-variation.

Proof. Let $(\mathcal{F}_n)_{n\geq 0}$ denote the filtration on J associated with the sequence of subdivisions. Since the mesh of $\mathcal{D}_n$ tends to 0, $\mathcal{F}_\infty = \sigma(\bigcup_{n\geq 0} \mathcal{F}_n)$ is the Borel σ-field of J. Let $\dot{X}$ be the derivative of X, which by assumption belongs to $L^1(J)$. For each $n \geq 0$, the derivative of $X(n)$ is $\dot{X}(n) = E[\dot{X}|\mathcal{F}_n]$. By the martingale convergence theorem, $\dot{X}(n)$ tends to $\dot{X}$ in $L^1(J)$ as n tends to ∞. The result follows because $\|X(n) - X\|_{1,J} = \|\dot{X}(n) - \dot{X}\|_{L^1(J)}$. $\qquad\square$

This proof uses the fact that the path takes its values in a finite-dimensional space. As a general rule, we will try not to use finite-dimensional arguments where they are not needed.

Let us consider now the general case and try to approximate elements of $\mathcal{V}^p(J, E)$ by piecewise linear paths. Given an element X of $C^0(J, E)$ and a partition $\mathcal{D}$ of J, let us denote by $X^{\mathcal{D}}$ the unique continuous path which coincides with X on the points of $\mathcal{D}$ and is affine on each of the sub-intervals of J delimited by $\mathcal{D}$.

Let X be a path with finite p-variation. The main result we are going to prove is that, as the mesh of $\mathcal{D}$ tends to 0, the paths $X^{\mathcal{D}}$ converge to X in q-variation for every $q > p$. We start by proving that the p-variation of $X^{\mathcal{D}}$ is not greater than that of X.

Lemma 1.12. *Let X be an element of $\mathcal{V}^p(J, E)$. Let $\mathcal{D}$ be a partition of J. Then*
$$\|X^{\mathcal{D}}\|_{p,J} \leq \|X\|_{p,J}.$$

Let us introduce the following notation. If X is a continuous path on J and $\mathcal{D} = (t_0, \ldots, t_r)$ is a partition of J, set
$$\|X\|_{p,\mathcal{D}} = \left[\sum_{i=0}^{r-1} |X_{t_i} - X_{t_{i+1}}|^p\right]^{\frac{1}{p}}.$$

Proof. Choose $\varepsilon > 0$. Let $\mathcal{D}_\varepsilon$ be a partition of J such that $\|X^{\mathcal{D}}\|_{p,\mathcal{D}_\varepsilon} \geq \|X^{\mathcal{D}}\|_{p,J} - \varepsilon$. We claim that $\mathcal{D}_\varepsilon$ can be chosen such that $\mathcal{D}_\varepsilon \subset \mathcal{D}$.

Indeed, assume that this inclusion does not hold. To start with, if $\mathcal{D}_\varepsilon$ does not contain the endpoints of J, then we can add them to $\mathcal{D}_\varepsilon$ and this can only increase $\|X^{\mathcal{D}}\|_{p,\mathcal{D}_\varepsilon}$. Now, assume that there is a time in $\mathcal{D}_\varepsilon$ which is not in $\mathcal{D}$ and let us consider the smallest such time, which we denote by u. Let t_i be the last time in $\mathcal{D}_\varepsilon$ before u, t_j the last time in $\mathcal{D}$ before u, and v the first time after u in $\mathcal{D} \cup \mathcal{D}_\varepsilon$. Since $s \mapsto X^{\mathcal{D}}_s$ is affine on $[t_j, v]$, the function $s \mapsto |X^{\mathcal{D}}_s - X^{\mathcal{D}}_{t_i}|^p + |X^{\mathcal{D}}_v - X^{\mathcal{D}}_s|^p$ is convex on $[t_j, v]$ and it must attain its maximum at one of the points t_j and v. Thus, by dropping u from $\mathcal{D}_\varepsilon$ and making sure that t_j or v, depending on where the function reaches its maximum, belongs to $\mathcal{D}_\varepsilon$, we do not decrease $\|X^{\mathcal{D}}\|_{p,\mathcal{D}_\varepsilon}$ but decrease by one the number of points in $\mathcal{D}_\varepsilon$ which are not in $\mathcal{D}$. By repeating this procedure a sufficient number of times, we can make sure that $\mathcal{D}_\varepsilon \subset \mathcal{D}$.

Since X and $X^{\mathcal{D}}$ coincide on $\mathcal{D}_\varepsilon$, it is now easy to conclude that

$$\|X^{\mathcal{D}}\|_{p,J} - \varepsilon \leq \|X^{\mathcal{D}}\|_{p,\mathcal{D}_\varepsilon} = \|X\|_{p,\mathcal{D}_\varepsilon} \leq \|X\|_{p,J}.$$

The result follows by letting ε tend to 0. $\qquad\square$

The next tool we need is a straightforward estimation of the distance in q-variation between two paths in function of their distance in p-variation when $q > p$.

Lemma 1.13. *Let p and q be such that $1 \leq p < q$. Let X, Y be two paths of $\mathcal{V}^p(J,E)$. Then*

$$\|X - Y\|_{\mathcal{V}^q(J,E)} \leq \left(2 \sup_{u \in J} |X_u - Y_u|\right)^{\frac{q-p}{q}} \|X - Y\|_{p,J}^{\frac{p}{q}} + \sup_{u \in J} |X_u - Y_u|.$$

If $\mathcal{D} = (t_0, \ldots, t_r)$ is a partition of $J = [0,T]$, let us denote its mesh by $|\mathcal{D}|$, i.e. set $|\mathcal{D}| = \max\{t_0, t_1 - t_0, \ldots, t_r - t_{r-1}, T - t_r\}$. We can now state the approximation result.

Proposition 1.14. *Let p and q be such that $1 \leq p < q$. Let X be a path of $\mathcal{V}^p(J,E)$. Then the paths $X^{\mathcal{D}}$ converge to X in q-variation norm as the mesh of $\mathcal{D}$ tends to 0. By this we mean that, for all $\varepsilon > 0$ there exists $\delta > 0$ such that, if $\mathcal{D}$ is a partition of J with $|\mathcal{D}| < \delta$, then $\|X^{\mathcal{D}} - X\|_{\mathcal{V}^q(J,E)} < \varepsilon$.*

In particular, if $(\mathcal{D}_n)_{n \geq 0}$ is a sequence of partitions of J whose mesh tends to 0 as n tends to ∞, the sequence of paths $(X^{\mathcal{D}_n})_{n \geq 0}$ converges to X in q-variation norm. One cannot expect the convergence to hold in p-variation norm, because this would imply that $\mathcal{V}^p(J,E)$ is separable.

Proof. Let $\mathcal{D}$ be a partition of J. By Lemma 1.13, we have

$$\|X^{\mathcal{D}} - X\|_{\mathcal{V}^q(J,E)} \leq \left(2 \sup_{u \in J} |X_u^{\mathcal{D}} - X_u|\right)^{\frac{q-p}{q}} \|X^{\mathcal{D}} - X\|_{p,J} + \sup_{u \in J} |X_u^{\mathcal{D}} - X_u|.$$

On one hand, since X is uniformly continuous on J, $\sup_{u \in J} |X_u^{\mathcal{D}} - X_u|$ can be made arbitrarily small by insisting that the mesh of $\mathcal{D}$ is small enough. On the other hand, Lemma 1.12 and the fact that $\|X^{\mathcal{D}} - X\|_{p,J}^p \leq 2^{p-1}(\|X^{\mathcal{D}}\|_{p,J}^p + \|X\|_{p,J}^p)$ imply that $\|X^{\mathcal{D}} - X\|_{p,J}$ is bounded uniformly with respect to $\mathcal{D}$. The result follows. $\qquad\square$

Lemma 1.13 has the following important consequence.

Proposition 1.15. *Assume that E is finite-dimensional. Let p and q be such that $1 \leq p < q$. Let $\mathcal{X}$ be a bounded subset of $\mathcal{V}^p(J,E)$. If $\mathcal{X}$ is uniformly equicontinuous, then it is relatively compact in $\mathcal{V}^q(J,E)$.*

In particular, if ω is a control on J, then the set of all paths with p-variation controlled by ω is relatively compact in $\mathcal{V}^q(J,E)$.

Proof. Assume that $\mathcal{X}$ is bounded in $\mathcal{V}^p$ and uniformly equicontinuous. Then $\mathcal{X}$ is relatively compact in the uniform topology. Hence, from every sequence in $\mathcal{X}$ one can extract a uniformly convergent sequence, which, by Lemma 1.13, converges in q-variation.

In particular, the set of paths whose p-variation is controlled by some fixed control ω is bounded in $\mathcal{V}^p$ and uniformly equicontinuous.   □

1.3 The Young Integral

Let X and Y be two real-valued functions on a compact interval $J = [0, T]$. For every partition $\mathcal{D} = (0 = t_0, \ldots, t_r = T)$ of J, let us introduce the notation

$$\int_{\mathcal{D}} Y \, dX = \sum_{i=0}^{r-1} Y_{t_i} (X_{t_{i+1}} - X_{t_i}).$$

For the sake of clarity, in this section we will only consider subdivisions that contain the endpoints of the intervals they subdivide.

With the example of the Stieltjes integral in mind, let us say that the integral $\int_0^T Y_s \, dX_s$ exists if $\int_{\mathcal{D}} Y \, dX$ converges as the mesh of $\mathcal{D}$ tends to 0, by which we mean that, for all $\varepsilon > 0$, there exists $\delta > 0$ such that, if $\mathcal{D}$ and $\mathcal{D}'$ are two subdivisions of $[0, T]$ such that $|\mathcal{D}| < \delta$ and $|\mathcal{D}'| < \delta$, then

$$\left| \int_{\mathcal{D}} Y \, dX - \int_{\mathcal{D}'} Y \, dX \right| < \varepsilon.$$

It is well known that if X has bounded variation, then the integral $\int_0^t Y \, dX$ exists for every $t \in J$. Moreover, as a function of t, this integral itself has bounded variation and, if $\| \cdot \|_{\infty, J}$ denotes the uniform norm on J, we have the inequality

$$\left\| \int_0^\bullet Y_s \, dX_s \right\|_{1, J} \leq \|Y\|_{\infty, J} \|X\|_{1, J}. \tag{1.2}$$

Young [9] discovered that in certain situations where X has unbounded variation, $\int Y \, dX$ still exists in the sense explained above. Of course, this requires more regularity on Y than just continuity. To be precise, he has shown that the integral exists when X has finite p-variation and Y has finite q-variation provided $\frac{1}{p} + \frac{1}{q} > 1$. The integral $\int_0^t Y_s \, dX_s$ thus defined is a continuous function of t with finite p-variation and it satisfies a relation similar to (1.2).

Our aim in this section is to explain Young's result. We do this for Banach-valued path, and this does not make the proof any more difficult. The main result is the following.

Theorem 1.16 (Young). *Let V and W be two Banach spaces. Let $p, q \geq 1$ be two real numbers such that $\frac{1}{p} + \frac{1}{q} > 1$. Let T be a positive real number. Consider $X \in \mathcal{V}^p([0,T], V)$ and $Y \in \mathcal{V}^q([0,T], \mathbf{L}(V,W))$. Then, for each $t \in [0,T]$, the limit*

$$\int_0^t Y_s \, dX_s = \lim_{|\mathcal{D}| \to 0, \mathcal{D} \subset [0,t]} \int_{\mathcal{D}} Y \, dX$$

exists. As a function of t, this limit belongs to $\mathcal{V}^p([0,T], W)$ and there exists a constant $C_{p,q}$ which depends only on p and q such that the following inequality holds:

$$\left\| \int_0^\bullet (Y_s - Y_0) \, dX_s \right\|_{p, [0,T]} \leq C_{p,q} \|Y\|_{q, [0,T]} \|X\|_{p, [0,T]}. \tag{1.3}$$

The main ingredient of this proof is a maximal inequality which provides us with a bound on $|\int_{\mathcal{D}} Y \, dX|$ uniform with respect to $\mathcal{D}$. We obtain this bound by successively *removing* points from $\mathcal{D}$. We will use this strategy again several times in these notes, but not anymore in such a relatively simple context.

Proof. Let us replace X and Y, respectively, by $\tilde{X}$ and $\tilde{Y}$, where for all $u \in [0,T]$,

$$\tilde{X}_u = \frac{X_u}{\|X\|_{p, [0,T]}}, \quad \tilde{Y}_u = \frac{Y_u}{\|Y\|_{q, [0,T]}}.$$

Let ω be the control defined for all $(u, v) \in \triangle_T$ by

$$\omega(u, v) = \frac{\|X\|_{p, [u,v]}^p}{\|X\|_{p, [0,T]}^p} + \frac{\|Y\|_{q, [u,v]}^q}{\|Y\|_{q, [0,T]}^q}.$$

Then ω satisfies $\omega(0, T) \leq 2$ and controls both the p-variation of $\tilde{X}$ and the q-variation of $\tilde{Y}$ on $[0, T]$.

Let $\mathcal{D} = (0 = t_0, \ldots, t_r = T)$ be a partition of $[0, T]$. As explained above, we want to bound $|\int_{\mathcal{D}} Y \, dX|$ independently of $\mathcal{D}$. For this, we remove a point of $\mathcal{D}$ in such a way that the variation of X and Y remains as evenly spread as possible between the different sub-intervals delimited by the partition. If $r = 2$, i.e. if $\mathcal{D}$ has only three points, let us choose the middle point t_1. If $r > 2$, let us choose t_i, where i is between 1 and $r - 1$ and such that

$$\omega(t_{i-1}, t_{i+1}) \leq \frac{2}{r - 1} \omega(0, T). \tag{1.4}$$

Such an i exists because, if the inequality was false for every $i = 1 \ldots r - 1$, we would have, by super-additivity, and assuming that r is even, $2\omega(0, T) < \omega(t_0, t_r) + \omega(t_1, t_{r-1}) \leq 2\omega(0, T)$. A similar contradiction would occur under the assumption that r is odd.

Observe that the inequality (1.4) holds even for $r = 2$. Now let us consider the partition $\mathcal{D} - \{t_i\}$. Using the notation $\tilde{X}^1_{u,v} = \tilde{X}_v - \tilde{X}_u$, we have

$$\left| \int_{\mathcal{D}} \tilde{Y} \, d\tilde{X} - \int_{\mathcal{D}-\{t_i\}} \tilde{Y} \, d\tilde{X} \right| = |\tilde{Y}_{t_{i-1}} \tilde{X}^1_{t_{i-1},t_i} + \tilde{Y}_{t_i} \tilde{X}^1_{t_i,t_{i+1}} - \tilde{Y}_{t_{i-1}} \tilde{X}^1_{t_{i-1},t_{i+1}}|$$

$$= |(\tilde{Y}_{t_i} - \tilde{Y}_{t_{i-1}})(\tilde{X}_{t_{i+1}} - \tilde{X}_{t_i})|$$

$$\leq \omega(t_{i-1}, t_i)^{\frac{1}{q}} \omega(t_i, t_{i+1})^{\frac{1}{p}}$$

$$\leq \omega(t_{i-1}, t_{i+1})^{\frac{1}{p}+\frac{1}{q}}$$

$$\leq \left(\frac{4}{r-1} \right)^{\frac{1}{p}+\frac{1}{q}}.$$

By iterating this procedure until the partition has only two points left, we find that

$$\left| \int_{\mathcal{D}} (Y - Y_s) \, dX \right| \leq 4^{\frac{1}{p}+\frac{1}{q}} \zeta \left(\frac{1}{p} + \frac{1}{q} \right) \|Y\|_{q,[0,T]} \|X\|_{p,[0,T]},$$

where ζ denotes the classical ζ function. Set $c_{p,q} = 4^{\frac{1}{p}+\frac{1}{q}} \zeta \left(\frac{1}{p} + \frac{1}{q} \right)$. Then we have established the following uniform bound:

$$\sup_{\mathcal{D} \subset [s,t]} \left| \int_{\mathcal{D}} Y \, dX \right| \leq c_{p,q} (\|Y\|_{q,[0,T]} + \|Y\|_{\infty,[0,T]}) \|X\|_{p,[0,T]}. \tag{1.5}$$

We have established this maximal inequality on $[0, T]$ but of course a similar inequality holds on every sub-interval of $[0, T]$. We will use this remark later.

We now use an approximation argument to show that the integral over $[0, T]$ exists. Let $p' > p$ be such that $\frac{1}{p'} + \frac{1}{q} > 1$. According to Proposition 1.14, there exists a sequence $(X(n))_{n \geq 0}$ of piecewise linear paths which converges in p'-variation to X. For each $n \geq 0$, $X(n)$ has bounded variation and

$$\lim_{|\mathcal{D}| \to 0, \mathcal{D} \subset [s,t]} \int_{\mathcal{D}} Y \, dX(n) = \int_s^t Y_u \, dX_u(n),$$

where the right-hand side is a Stieltjes integral. On the other hand, the uniform bound (1.5) gives

$$\sup_{\mathcal{D} \subset [0,T]} \left| \int_{\mathcal{D}} Y \, dX(n) - \int_{\mathcal{D}} Y \, dX \right|$$

$$\leq c_{p',q} (\|Y\|_{q,[0,T]} + \|Y\|_{\infty,[0,T]}) \|X(n) - X\|_{p',[0,T]}.$$

Hence, this supremum tends to 0 as n tends to ∞. It is now an elementary exercise to combine the last two equations in order to prove that

$$\lim_{\varepsilon \to 0} \sup_{|\mathcal{D}| \vee |\mathcal{D}'| < \varepsilon} \left| \int_{\mathcal{D}} Y \, dX - \int_{\mathcal{D}'} Y \, dX \right| = 0.$$

Hence, $\int_0^T Y_u \, dX_u$ exists as the limit of $\int_{\mathcal{D}} Y \, dX$ when the mesh of $\mathcal{D}$ tends to 0.

For every $(s,t) \in \triangle_T$, the same proof applied to the sub-interval $[s,t]$ of $[0,T]$ shows that the integral $\int_s^t Y_u \, dX_u$ exists. According to Lemma 1.10, the maximal inequalities corresponding to (1.5) on every sub-interval of $[0,T]$ imply that the path $t \mapsto \int_0^t Y_s \, dX_s$ has finite p-variation. If $Y_0 = 0$, then $\|Y\|_{\infty, J} \leq \|Y\|_{q,J}$, so that the inequality (1.3) holds with $C_{p,q} = 2c_{p,q}$. $\square$

Remark 1.17. If the p-variation of X is controlled by some ω, then the maximal inequality (1.5) on $[0,T]$ and its sub-intervals implies that the p-variation of $\int_0^\bullet Y_s \, dX_s$ is controlled by $c_{p,q}^p \|Y\|_{\mathcal{V}^q([0,T])}^p \omega$. Moreover, (1.5) also implies that $\|\int_0^\bullet Y_s \, dX_s\|_{\infty, [0,T]} \leq c_{p,q} \|Y\|_{\mathcal{V}^q([0,T])} \|X\|_{p,[0,T]}$. Since $C_{p,q} = 2c_{p,q}$, we find in particular that

$$\left\| \int_0^\bullet Y_s \, dX_s \right\|_{\mathcal{V}^p([0,T],W)} \leq C_{p,q} \|Y\|_{\mathcal{V}^q([0,T],W)} \|X\|_{p,[0,T]}.$$

1.4 Differential Equations Driven by Signals with Finite p-Variation, with $p < 2$

1.4.1 Peano's Theorem

When X has bounded variation, Theorem 1.4 asserts that the differential equation $dY_t = f(Y_t) \, dX_t$ admits solutions as soon as f is continuous. If we expect to solve differential equations driven by more irregular paths, the function f will presumably need to be smoother. For example, if we take X with finite p-variation for some $p \geq 1$, then any solution Y must have finite p-variation and, to make sense of the integral $\int f(Y) \, dX$, we need $f(Y)$ to have finite q-variation for some q such that $\frac{1}{p} + \frac{1}{q} > 1$. Not every continuous f does that, but it is easy to identify a class of vector fields which will do.

As usual, V and W are two Banach spaces. The signal X is a continuous V-valued path and the vector field is a continuous mapping $f : W \to \mathbf{L}(V,W)$.

Lemma 1.18. *Assume that f is Hölder continuous of exponent γ with $0 < \gamma \leq 1$. Assume also that $Y \in \mathcal{V}^p(J,W)$ for some $p \geq 1$. Then $f \circ Y$ belongs to $\mathcal{V}^{\frac{p}{\gamma}}(J, \mathbf{L}(V,W))$ and*

$$\|f \circ Y\|_{\mathcal{V}^{\frac{p}{\gamma}}} \leq K \|Y\|_{\mathcal{V}^p}^\gamma.$$

Proof. Assume that $K > 0$ is such that, for all $w, w' \in W$, $|f(w) - f(w')| \leq K|w - w'|^\gamma$. Then, for every partition $\mathcal{D} \subset J$,

$$\|f \circ Y\|_{\frac{p}{\gamma}, \mathcal{D}} \leq K \|Y\|_{p, \mathcal{D}}^\gamma.$$

The result follows immediately. $\square$

Remark 1.19. It is a hard problem to study the continuity of the map $Y \mapsto f(Y)$. What can easily be said is the following: if a sequence $(Y_n)_{n \geq 0}$ converges in p-variation norm to Y, then the sequence $(f(Y_n))_{n \geq 0}$ is bounded in $\frac{p}{\gamma}$-variation and converges uniformly to $f(Y)$. Hence, by Lemma 1.13, it converges to $f(Y)$ in $\frac{p'}{\gamma}$-variation norm for every $p' > p$.

If X has finite p-variation, the condition on f for our differential equation to make sense is thus $\frac{1}{p} + \frac{\gamma}{p} > 1$, i.e. $\gamma > p - 1$. In particular, this requires that $p < 2$, since $\gamma \leq 1$. Let us use the notation $f \in \mathrm{Lip}(\gamma)$ to indicate that f is Hölder continuous with exponent γ.

Theorem 1.20 (Peano). *Assume that W is finite-dimensional. Let p and γ be such that $1 \leq p < 2$ and $p - 1 < \gamma \leq 1$. Assume that X has finite p-variation and that f is $\mathrm{Lip}(\gamma)$. Then, for every $\xi \in W$, the differential equation (1.1) admits a solution.*

The proof is similar to that of Theorem 1.4 and we give it here with more details.

Proof. Assume that $J = [0, T]$ and choose $t \in J$. Let us also choose $\xi \in W$. Consider $p'' > p' > p$ such that $\gamma > p'' - 1$. By Theorem 1.16 and Lemma 1.18, one can define a functional $F : \mathcal{V}^{p'}([0, t], W) \longrightarrow \mathcal{V}^{p'}([0, t], W)$ by

$$F(Y_\bullet) = \xi + \int_0^\bullet f(Y_s) \, dX_s.$$

Assume that a sequence $(Y_n)_{n \geq 0}$ converges in p'-variation norm to Y. Then, by the Remark 1.19, $f(Y_n)$ tends to $f(Y)$ in $\frac{p''}{\gamma}$-variation norm and, by Theorem 1.16 again, $F(Y_n)$ tends to $F(Y)$ in p-variation norm, hence in p'-variation norm. We have proved that the functional F is continuous.

Recall that $\mathcal{V}^{p'}([0, t], W)$ is a Banach space when it is endowed with the norm $\|\cdot\|_{\mathcal{V}^{p'}}$. Set $M = \max(1, 2|\xi|)$ and assume that $\|Y\|_{\mathcal{V}^{p'}} \leq M$. Finally, let K be the Hölder constant of f. Then, by the Remark 1.17 and Lemma 1.18,

$$\|F(Y)\|_{\mathcal{V}^{p'}} \leq |\xi| + \left\| \int_0^\bullet f(Y_s) \, dX_s \right\|_{\mathcal{V}^{p'}}$$

$$\leq |\xi| + C_{p, \frac{p'}{\gamma}} K \|Y\|_{\mathcal{V}^{p'}}^\gamma \|X\|_{p, [0, t]}$$

$$\leq M \left(\frac{1}{2} + C_{p, \frac{p'}{\gamma}} K \|X\|_{p, [0, t]} \right).$$

The last inequality shows that, by choosing t small enough, we may enforce that the ball $B = \{Y \in \mathcal{V}^{p'}([0, t], W) \mid \|Y\|_{\mathcal{V}^{p'}} \leq M\}$ is stabilised by F. Let us choose t in that way.

On the other hand, assume that the p-variation of X is controlled by ω. Then the Remark 1.17 shows that the elements of $F(B)$ have p-variation uniformly controlled by $C\omega$ for some constant C. By Proposition 1.15, this shows that $F(B)$ is compact in $\mathcal{V}^{p'}([0,t],W)$.

Schauder's theorem applies now to ensure that F has a fixed point in B. This fixed point is a solution of (1.1) on $[0,t]$. A solution on $[0,T]$ can now easily be obtained by subdividing the interval $[0,T]$ in finitely many subintervals on which the p-variation of X is small enough for the argument above to apply. $\qquad\square$

At the beginning of this proof, it would have been much easier to prove the continuity of F if we would have been able to say that $f(Y)$ depends continuously on Y in $\frac{p}{\gamma}$-variation norm. Fortunately, approximation came to our rescue. If we now seek assumptions under which there is uniqueness of the solution, we need to have a much better control on the mapping $Y \mapsto f(Y)$. This requires f to be more regular than Hölder continuous. We introduce now the right spaces of one-forms.

1.4.2 Lipschitz Functions

Let V be, as usual, a Banach space. We are going to define a notion of smooth functions which makes sense for functions defined on rather small sets of V, typically closed sets. For the sake of simplicity, let us assume that the dimension of V is finite: this makes the definition of the tensor powers of V easier. We make general remarks on tensor products and on the infinite-dimensional case after the main definition.

Let us explain the main idea on polynomial functions first. Let $k \geq 0$ be an integer. Let $P : V \longrightarrow \mathbb{R}$ be a polynomial of degree k. Let $X : [0,T] \longrightarrow V$ be a Lipschitz continuous path. Then, if $P^1 : V \longrightarrow \mathbf{L}(V,\mathbb{R})$ denotes the derivative of P, then Taylor's theorem asserts that, for all $t \in [0,T]$,

$$P(X_t) = P(X_0) + \int_0^t P^1(X_s)\,dX_s.$$

Let now $P^2 : V \longrightarrow \mathbf{L}(V \otimes V, \mathbb{R})$ denote the second derivative of P. Recall that $\mathbf{L}(V \otimes V, \mathbb{R})$ is the space of bilinear forms on V. Then, by substitution into the last expression, we get

$$P(X_t) = P(X_0) + P^1(X_0)\int_{0<u_1<t} dX_{u_1} + \iint_{0<u_1<u_2<t} P^2(X_{u_1})\,dX_{u_1}\otimes dX_{u_2}.$$

If we keep substituting, the procedure stops when we reach the $(k+1)$th derivative of P. At that time, we get the expression

$$P(X_t) = P(X_0) + P^1(X_0)\int_{0<u_1<t} dX_{u_1} + P^2(X_0)\iint_{0<u_1<u_2<t} dX_{u_1}\otimes dX_{u_2} + \dots$$

$$+ P^k(X_0)\int_{0<u_1<\dots<u_k<t}\cdots\int dX_{u_1}\otimes\dots\otimes dX_{u_k}. \tag{1.6}$$

For each $j \in \{1, \dots, k\}$, $P^j : V \longrightarrow \mathbf{L}(V^{\otimes j}, \mathbb{R})$ is the jth derivative of P and takes in fact its values in the space of *symmetric j-linear forms* on V. Hence, for each j, $P^j(X_0) \int dX_{u_1} \otimes \dots \otimes dX_{u_j}$ depends only on the symmetric part of the tensor $\int dX_{u_1} \otimes \dots \otimes dX_{u_j}$, which is equal to

$$\frac{1}{j!} \sum_{\sigma \in \mathfrak{S}_j} \int_{0 < u_1 < \dots < u_j < t} dX_{u_{\sigma(1)}} \otimes \dots \otimes dX_{u_{\sigma(n)}} = \frac{1}{j!}(X_t - X_0)^{\otimes j}.$$

Equation (1.6) says nothing more about P than the fact that

$$P(X_t) = P(X_s) + \sum_{j=1}^{k} P^j(X_s)\frac{(X_t - X_s)^{\otimes j}}{j!}. \tag{1.7}$$

In general, if $k \geq 0$ is an integer and $\gamma \in (k, k+1]$, we shall say that a function f is Lip(γ) if there exists k functions $f^1, \dots, f^k$ taking their values in the spaces of symmetric multi-linear forms such that a relation similar to (1.6) – or (1.7), as it is the same – holds, up to an error which is Hölder continuous of exponent $\gamma - k$.

Definition 1.21. *Let V and W be two Banach spaces. Let $k \geq 0$ be an integer. Let $\gamma \in (k, k+1]$ be a real number. Let F be a closed subset of V. Let $f : F \longrightarrow W$ be a function. For each integer $j = 1, \dots, k$, let $f^j : F \longrightarrow \mathbf{L}(V^{\otimes j}, W)$ be a function which takes its values in the space of symmetric j-linear mappings from V to W. The collection $(f = f^0, f^1, \dots, f^k)$ is an element of Lip(γ, F) if the following condition holds.*

There exists a constant M such that, for each $j = 0, \dots, k$,

$$\sup_{x \in F} |f^j(x)| \leq M$$

and there exists a function $R_j : V \times V \longrightarrow \mathbf{L}(V^{\otimes j}, W)$ such that, for each $x, y \in F$ and each $v \in V^{\otimes j}$,

$$f^j(y)(v) = \sum_{l=0}^{k-j} \frac{1}{l!} f^{j+l}(x)(v \otimes (y - x)^{\otimes l}) + R_j(x, y)(v)$$

and

$$|R_j(x, y)| \leq M|x - y|^{\gamma - j}.$$

We usually say that f is Lip(γ, F) without mentioning explicitly $f^1, \dots, f^k$. The smallest constant M for which the inequalities hold for all j is called the Lip(γ, F)-norm of f and it is denoted by $\|f\|_{\mathrm{Lip}(\gamma)}$.

A few remarks about this definition are in order.

Remark 1.22. Assume that $(f, f^1, \dots, f^k)$ belongs to Lip(γ, F). Because the functions $f^1, \dots, f^k$ are required to take their values in the space of symmetric multi-linear functions from V to W, we have for each Lipschitz continuous path $X : [0,T] \longrightarrow F$, each $0 \le s < t \le T$, each $j = 1,\ldots,k$ and each $v \in V^{\otimes j}$ the relation

$$f^j(X_t)(v) = \sum_{l=0}^{k-j} f^{j+l}(X_s)\left(v \otimes \int_{s<u_1<\ldots<u_l<t} dX_{u_1} \otimes \ldots \otimes dX_{u_l} \right)$$
$$+ R_j(X_s, X_t)(v),$$

which is reminiscent of, and in fact equivalent to, (1.6). It is in general in this form that we use the assumption that f is $\mathrm{Lip}(\gamma, F)$.

Remark 1.23. If $(f, f^1, \ldots, f^k)$ belongs to $\mathrm{Lip}(\gamma, F)$, then the functions $f^1, \ldots, f^k$ may not be uniquely determined by f. On the interior of F, they are the classical derivatives of F, but this interior may be empty. In general they are only polynomial approximations of f at increasing orders and, if F is contained for example in a hyperplane, then the functions $f^1, \ldots, f^k$ are not determined by f in the directions transverse to this hyperplane.

On the other hand, it is a classical result of Whitney (see [80, Sect. VI.2]) that, under the assumption that V has finite dimension, there exists, for all closed F, a continuous extension operator $\mathrm{Lip}(\gamma, F) \longrightarrow \mathrm{Lip}(\gamma, V)$. Thus, given a function f defined on F, there exist $f^1, \ldots, f^k$ such that $(f, f^1, \ldots, f^k)$ belongs to $\mathrm{Lip}(\gamma, F)$ if and only if f can be extended to an element $\mathrm{Lip}(\gamma, V)$, i.e. a bounded function on V which is k times continuously differentiable with bounded derivatives and whose kth differential is Hölder continuous with exponent $\gamma - k$. Unfortunately, no such results holds in infinite dimensions.

Remark 1.24. In our definition of Lip functions, we have used norms on the tensor powers of V without specifying them explicitly. As long as V is finite dimensional, this does not matter as different tensor norms produce equivalent Lip-norms. However, if V is infinite dimensional, this is not the case anymore.

In order to understand this fairly concretely, it is useful to think of the tensor powers of V as spaces of homogeneous non-commuting polynomials in a family of variables indexed by a basis of V. Let $(v_1, \ldots, v_n)$ be a basis of V, which for the moment we assume to be finite dimensional. Let $r \ge 1$ be an integer. Then a basis of $V^{\otimes r}$ is given by the set of tensors $v_I = v_{i_1} \otimes \ldots \otimes v_{i_r}$, where $I = (i_1, \ldots, i_r)$ spans $\{1, \ldots, n\}^r$. Thus, if $(\alpha_I)_{I \in \{1,\ldots,n\}^r}$ is a collection of real numbers, then the tensor $\sum_I \alpha_I v_I$ can be identified with the polynomial $\sum_I \alpha_I X_I$, where $X_1, \ldots, X_n$ are indeterminates and $X_I = X_{i_1} \ldots X_{i_r}$. This identification between tensors of finite maximal degree and non-commuting polynomials will be extremely helpful in Chap. 2.

For the moment, the basis of $V^{\otimes r}$ we have just written gives rise to several obvious norms, like $|\sum_I \alpha_I X_I|_1 = \sum_I |\alpha_I|$ or $|\sum_I \alpha_I X_I|_\infty = \sup_I |\alpha_I|$. If V is a Hilbert space, then it is certainly natural to choose an orthonormal basis of V and to choose the norm $|\sum_I \alpha_I X_I|_2 = (\sum_I |\alpha_I|^2)^{\frac{1}{2}}$.

The symmetric group of order r acts on $V^{\otimes r}$ by permuting the letters and all the norms which we have considered are preserved by this action. This is a property that we will always assume to be fulfilled, even when V is infinite dimensional. Even with this restriction, there are several non-equivalent natural norms on $V^{\otimes r}$. If V is a Hilbert space, there is a canonical choice which is a Hilbert norm, namely the $|\cdot|_2$ norm written above. If V is only a Banach space, there are two norms which play a particular role, namely the projective and the injective norm, corresponding, respectively, to the norms $|\cdot|_1$ and $|\cdot|_\infty$ and which are in a sense, respectively, the largest and the smallest sensible norms on $V^{\otimes r}$.

Once a norm is fixed on the tensor powers of V, the definition of Lip functions makes perfect sense even for infinite-dimensional spaces.

For further reference, let us state the properties of the norms on tensor products which we assume to be true.

Definition 1.25. *Let V be a Banach space. We say that its tensor powers are endowed with admissible norms if the following conditions hold:*

1. For each $n \geq 1$, the symmetric group S_n acts by isometries on $V^{\otimes n}$, i.e.

$$\|\sigma v\| = \|v\| \quad \forall v \in V^{\otimes n}, \ \forall \sigma \in S_n.$$

2. The tensor product has norm 1, i.e. for all $n, m \geq 1$,

$$\|v \otimes w\| \leq \|v\|\|w\| \quad \forall v \in V^{\otimes n}, \ w \in V^{\otimes m}.$$

From now on, we consider only vector fields f which belong to $\mathrm{Lip}(\gamma, V)$ for some γ. The following property is essential in understanding the mapping $Y \mapsto f(Y)$.

Proposition 1.26 (Division Property). *Take $\gamma > 1$. Let f be a $\mathrm{Lip}(\gamma)$ function on V with values in W. Then there exists a function $g : V \times V \longrightarrow \mathbf{L}(V, W)$ which is $\mathrm{Lip}(\gamma - 1)$ and such that, for all x and y in V, one has*

$$f(x) - f(y) = g(x, y)(x - y).$$

Moreover, there exists a constant C which depends only on γ and V such that

$$\|g\|_{\mathrm{Lip}(\gamma-1)} \leq C\|f\|_{\mathrm{Lip}(\gamma)}.$$

Proof. Let k be the integer such that $k < \gamma \leq k + 1$. Let $f^1, \ldots, f^k$ be the k first differentials of f. Let also $R_0, \ldots, R_k$ be the error terms in the Taylor expansions of $f, f^1, \ldots, f^k$. For all $x, y \in V$, set

$$g(x, y) = \int_0^1 f^1(tx + (1 - t)y) \, dt \in \mathbf{L}(V, W)$$

and, for all $j = 0, \ldots, k - 1$ and $u, v \in V$, set

$$g^j(x,y)(u,v)^{\otimes j} = \int_0^1 f^{j+1}(tx + (1-t)y)(tu + (1-t)v)^{\otimes j} \; dt \in \mathbf{L}(V,W).$$

Finally, for all $x', y', u, v \in V$, set

$$S_j((x,y),(x',y'))(u,v)^{\otimes j} = \int_0^1 R_{j+1}((tx + (1-t)y),(tx' + (1-t)y'))$$
$$\times (tu + (1-t)v)^{\otimes j} \; dt.$$

Then it is left as an exercise for the reader to check that $(g^0, \dots, g^{k-1})$ is a $\mathrm{Lip}(\gamma - 1)$ function on $V \times V$ with error terms $S_0, \dots, S_{k-1}$. Moreover, $g = g^0$ satisfies the relation $f(x) - f(y) = g(x,y)(x-y)$ for all $x, y \in V$. Finally, let C_V be an upper bound for the Lipschitz norms of the Lipschitz continuous mappings $(x,y) \mapsto tx + (1-t)y$ from $V \times V$ to V, as t runs through $[0,1]$. Then

$$\|g\|_{\mathrm{Lip}(\gamma-1)} \leq \max_{j=1\dots k} C_V^{\gamma - j} \|f\|_{\mathrm{Lip}(\gamma)}.$$

The result is proved. $\hfill\square$

We can now prove a regularity result for the mapping $Y \mapsto f(Y)$. Let W and U be two Banach spaces.

Proposition 1.27. *Assume that* $f : W \longrightarrow U$ *is* $\mathrm{Lip}(1 + \alpha)$ *for some* $\alpha \in (0,1]$. *Let* $p \geq 1$ *be fixed. For every* $K > 0$, *there exists* $C_{\alpha,K} > 0$ *such that, whenever* X *and* Y *are two paths of* $\mathcal{V}^p(J, W)$ *such that* $\|X\|_{\mathcal{V}^p} \leq K$ *and* $\|Y\|_{\mathcal{V}^p} \leq K$, *one has*

$$\|f(X) - f(Y)\|_{\mathcal{V}^{\frac{p}{\alpha}}} \leq C_{\alpha,K} \|f\|_{\mathrm{Lip}(1+\alpha)} \|X - Y\|_{\mathcal{V}^p}.$$

Proof. Let $g : W \times W \longrightarrow \mathbf{L}(W, U)$ be a $\mathrm{Lip}(\alpha)$ function such that, for all $x, y \in W$, $f(x) - f(y) = g(x,y)(x-y)$. Pick $s, t \in J$. Then

$$|(f(X_t) - f(Y_t)) - (f(X_s) - f(Y_s))|^{\frac{p}{\alpha}}$$
$$= |g(X_t, Y_t)(X_t - Y_t) - g(X_s, Y_s)(X_s - Y_s)|^{\frac{p}{\alpha}}$$
$$= |g(X_t, Y_t)((X_t - Y_t) - (X_s - Y_s)) + (g(X_t, Y_t) - g(X_s, Y_s))(X_s - Y_s)|^{\frac{p}{\alpha}}$$
$$\leq 2^{\frac{p}{\alpha} - 1} |g(X_t, Y_t)|^{\frac{p}{\alpha}} |(X_t - Y_t) - (X_s - Y_s)|^{\frac{p}{\alpha}} +$$
$$+ 2^{\frac{p}{\alpha} - 1} \|g\|_{\mathrm{Lip}(\alpha)}^{\frac{p}{\alpha}} |(X_t, Y_t) - (X_s, Y_s)|^p |X_s - Y_s|^{\frac{p}{\alpha}}.$$

Let C be a constant such that, for all $x, y, x', y' \in V$, $|(x,y) - (x',y')| \leq C(|x - x'| + |y - y'|)$. Then from the inequality above, we deduce the following:

$$\|f(X) - f(Y)\|_{\frac{p}{\alpha}, J}^{\frac{p}{\alpha}} \leq 2^{\frac{p}{\alpha} - 1} \sup_{t \in J} |g(X_t, Y_t)|^{\frac{p}{\alpha}} \|X - Y\|_{\frac{p}{\alpha}, J}^{\frac{p}{\alpha}}$$
$$+ 2^{\frac{p}{\alpha} - 1} \|g\|_{\mathrm{Lip}(\alpha)}^{\frac{p}{\alpha}} C^p 2^{p-1} (\|X\|_{p,J}^p + \|Y\|_{p,J}^p) \|X - Y\|_{\infty, J}^{\frac{p}{\alpha}}.$$

The result follows easily. $\hfill\square$

1.4.3 Picard's Theorem

It is now possible to prove a theorem similar to Picard's theorem for (1.1). Just as in the classical situation where X has bounded variation, the proof of the uniqueness of the solution requires exactly one more degree of regularity on f than the proof of its existence.

Theorem 1.28 (Picard). *Let p and γ be such that $1 \leq p < 2$ and $p < \gamma$. Assume that X has finite p-variation and that f is $\mathrm{Lip}(\gamma)$. Then, for every $\xi \in W$, the differential equation (1.1) admits a unique solution.*

Let $Y = I_f(X, \xi)$ denote the unique solution to (1.1) starting at ξ. Then the mapping $I_f : \mathcal{V}^p(J, V) \times W \longrightarrow \mathcal{V}^p(J, W)$ is continuous.

Proof. Let $\alpha \in (0, 1]$ be such that $\gamma \geq 1+\alpha > p$, so that f is $\mathrm{Lip}(1+\alpha)$. Notice that $p < 1 + \alpha$ is equivalent to $\frac{1}{p} + \frac{\alpha}{p} > 1$. Pick $\xi \in W$, $t \in J = [0, T]$ and consider as usual the functional $F : \mathcal{V}^p([0,t], W) \longrightarrow \mathcal{V}^p([0,t], W)$ defined by

$$F(Y_\bullet) = \xi + \int_0^\bullet f(Y_s)\, dX_s.$$

Choose $M = 2|\xi|$. Let Y in $\mathcal{V}^p([0,t], W)$ be such that $\|Y\|_{\mathcal{V}^p} \leq M$. Since f is in particular $\mathrm{Lip}(1)$, i.e. Lipschitz continuous, $f(Y)$ has finite p-variation and is controlled by $\|f\|_{\mathrm{Lip}(1)}$ times that of Y. Hence, by Theorem 1.16 and Remark 1.17,

$$\begin{aligned}
\|F(Y)\|_{\mathcal{V}^p([0,t], W)} &\leq |\xi| + 2 \left\| \int_0^\bullet f(Y_s)\, dX_s \right\|_{p, [0,t]} \\
&\leq |\xi| + 2 C_{p,p} \|f\|_{\mathrm{Lip}(1)} \|Y\|_{p,[0,t]} \|X\|_{p,[0,t]} \\
&\leq M \left(\frac{1}{2} + C_{p,p} \|f\|_{\mathrm{Lip}(1)} \|X\|_{p,[0,t]} \right).
\end{aligned} \tag{1.8}$$

On the other hand, let Y and Y' be two elements of $\mathcal{V}^p([0,t], W)$ such that $\|Y\|_{\mathcal{V}^p} \leq M$ and $\|Y'\|_{\mathcal{V}^p} \leq M$. By Proposition 1.27,

$$\|f(Y) - f(Y')\|_{\mathcal{V}^{\frac{p}{\alpha}}} \leq C_{\alpha, M} \|f\|_{\mathrm{Lip}(1+\alpha)} \|Y - Y'\|_{\mathcal{V}^p},$$

so that, by Remark 1.17,

$$\|F(Y) - F(Y')\|_{\mathcal{V}^p} \leq C_{p, \frac{p}{\alpha}} C_{\alpha, M} \|f\|_{\mathrm{Lip}(1+\alpha)} \|Y - Y'\|_{\mathcal{V}^p} \|X\|_{p,[0,t]}. \tag{1.9}$$

If t is chosen small enough for the inequality

$$\|X\|_{p,[0,t]} < C_{p, \alpha, M, f} = \left(2 C_{p,p} \|f\|_{\mathrm{Lip}(1)}\right)^{-1} \wedge \left(2 C_{p, \frac{p}{\alpha}} C_{\alpha, M} \|f\|_{\mathrm{Lip}(1+\alpha)}\right)^{-1} \tag{1.10}$$

to hold, then the inequalities (1.8) and (1.9) tell us that F stabilises the ball $\{\|\cdot\|_{\mathcal{V}^p([0,t], W)} \leq M\}$ and is contractant on this ball. Hence, F admits a fixed point on this ball, which is the unique solution to (1.1) on the interval $[0, t]$.

The existence and uniqueness of the solution on the whole interval $J = [0, T]$ follow as always by subdividing it into shorter sub-intervals where the p-variation of X is small enough for the argument above to apply.

Let $Y = I_f(X, \xi)$ denote the unique solution to (1.1) starting at ξ. Choose X in $\mathcal{V}^p([0, T], V)$, $M > 0$ and let t be such that (1.10) holds. Then, it follows immediately from (1.9) that, if $|\xi| \leq \frac{M}{2}$ and $\|Y\|_{\mathcal{V}^p([0,t])} \leq M$, then for all $n \geq 1$,

$$\|F^n(Y) - I_f(X, \xi)\|_{\mathcal{V}^p([0,t])} \leq \frac{M}{2^{n-2}},$$

where F^n is the nth iteration $F \circ \ldots \circ F$ of F. Thus, the sequence of continuous mappings $(\xi, X, Y) \mapsto (\xi, X, F^n(Y))$ converges uniformly to the mapping $(\xi, X, Y) \mapsto (\xi, X, I_f(X, \xi))$ on the domain $\{(\xi, X, Y) \mid 2|\xi| \leq M, \|X\|_{p,[0,t]} < C_{p,\alpha,M,f}, \|Y\|_{\mathcal{V}^p([0,t])} \leq M\}$. The latter mapping is thus continuous on this domain, and so is in particular the Itô functional

$$I_f : \{(X, \xi) \mid \|X\|_{p,[0,t]} \leq C_{p,\alpha,M,f}, 2|\xi| \leq M\} \longrightarrow \mathcal{V}^p([0,t], W).$$

It is left to the reader to check that the continuity holds actually on the whole space $\mathcal{V}^p([0,t], V) \times W$. $\square$

1.5 What Goes Wrong with $p \geq 2$

In this chapter, we have described an extension of the classical theory of ordinary differential equations to some cases where the signal has unbounded variation. Our main tool has been the Young integral, which makes sense of the expression $\int Y \, dX$ in some cases where it is not in the scope of measure theory. The main limitation of the above theory seems to lie in the constraint $\frac{1}{p} + \frac{1}{q} > 1$ which must hold on the regularities of X and Y for the Young integral $\int Y \, dX$ to exist. Hence, even if the vector field f is Lipschitz continuous, so that $f(Y)$ is not less regular than Y on the scale of p-variation, we are not yet even able to give a meaning to (1.1) when the signal X has finite 2-variation but infinite p-variation for every $p < 2$. With probabilistic applications in mind, this is a drastic restriction, since Brownian paths have finite p-variation only for $p > 2$.

We claim that the threshold at $p = 2$ is not an artefact due to some particular limitation of the Young integral or any other weakness of our approach in this chapter, but rather that it has a deep significance and that the case $p \geq 2$ requires an essentially more sophisticated treatment than the case $p < 2$. In this paragraph, we give two important hints at the essential inadequacy of our present point of view for the case $p \geq 2$.

1.5.1 No Continuous Extension of the Stieltjes Integral is Rich Enough to Handle Brownian Paths

Let B be a Banach space of real-valued 1-periodic functions on $\mathbb{R}$. For each $t \in \mathbb{R}$ and each positive n, set $c_n(t) = \cos(2\pi nt)$ and $s_n(t) = \sin(2\pi nt)$.

Let $(C_n)_{n\geq 1}$ and $(S_n)_{n\geq 1}$ be two sequence of independent standard Gaussian random variables. We say that B *carries the Wiener measure* if, for all $n \geq 1$, c_n and s_n belong to B, and if the series

$$\sum_{n=1}^{\infty} \frac{C_n c_n + S_n s_n}{2\pi n}$$

converges almost surely in B.

Proposition 1.29. *Let B be a Banach space which carries the Wiener measure. There exists no bilinear continuous functional $I : B \times B \longrightarrow \mathbb{R}$ such that, if $x, y \in B$ are trigonometric polynomials, then $I(y, x) = \int_0^1 y_t \, dx_t$.*

Proof. Assume such a continuous bilinear form I exists. For each $N \geq 1$, define

$$W_N = \sum_{n=1}^{N} \frac{C_n c_n + S_n s_n}{2\pi n}, \quad \widetilde{W}_N = \sum_{n=1}^{N} \frac{C_n s_n - S_n c_n}{2\pi n}.$$

By assumption, W_N and $\widetilde{W}_N$ converge almost surely as N tends to ∞, respectively, to W and $\widetilde{W}$, two 1-periodic Brownian motions.

Now, by assumption on I, we have for all $N \geq 1$

$$I(W_N, \widetilde{W}_N) = \int_0^1 W_N \, d\widetilde{W}_N = \sum_{n=1}^{N} \frac{C_n^2 + S_n^2}{2\pi n}.$$

The last sum tends on one hand to $I(W, \widetilde{W})$ and on the other hand, it diverges almost surely to $+\infty$ as N tends to ∞. This contradiction terminates the proof. $\qquad\square$

This counter-example shows that there cannot exist a linear theory of integration which is rich enough to handle Brownian paths in a deterministic fashion.

1.5.2 The Area Enclosed by a Path is Not a Continuous Functional in Two-Variation

Consider (1.1) in the case where $W = V \oplus (V \otimes V)$ and $f : W \longrightarrow \mathbf{L}(V, W)$ is defined as follows: if $(y^1, y^2) \in W$ and $x \in V$, then

$$f(y^1, y^2)x = (x, y^1 \otimes x).$$

This definition is more natural than it may look: if X is a Lipschitz continuous V-valued path, then the unique solution to (1.1) starting at $(0,0)$ with this definition of f is easily checked to be

$$Y_t = (Y_t^1, Y_t^2) = \left(\int_{0 < u_1 < t} dX_{u_1}, \int_{0 < u_1 < u_2 < t} dX_{u_1} \otimes dX_{u_2} \right).$$

The point we want to make is that the second iterated integral of a path is a solution to a differential equation driven by this path, and with a polynomial one-form. However demanding conditions we are prepared to put on the one-form in order to allow more irregular signals to drive our equation, polynomial one-forms could hardly be left aside. Then, if there would exist for $p = 2$ a theorem of existence and uniqueness of the solution with continuous dependence on the signal like Theorem 1.28, then the second iterated integral of a Lipschitz continuous path would be a continuous functional of this path in the topology two-variation. We show that this is not true if V is at least of dimension 2.

Proposition 1.30. *The mapping* $X \mapsto \int_{0<u_1<u_2<1} dX_{u_1} \otimes dX_{u_2}$ *defined on* $\mathcal{V}^1([0,1], \mathbb{R}^2)$ *is not continuous with respect to the norm* $\|\cdot\|_{2,[0,1]} + \|\cdot\|_{\infty,[0,1]}$.

In fact, we prove a slightly weaker statement, namely that the mapping is not continuous in any $\|\cdot\|_{\mathcal{V}^p}$-norm for $p > 2$. The reader is invited to adapt the proof in the case $p = 2$.

Before we give the counter-example, let us discuss the information contained in the second iterated integral of a Lipschitz continuous path. The tensor $\int_{0<u_1<u_2<t} dX_{u_1} \otimes dX_{u_2}$ belongs to $V \otimes V$ and can be decomposed into its symmetric and antisymmetric parts. Its symmetric part is simply

$$S_t = \frac{1}{2} \int_{0<u_1,u_2<t} dX_{u_1} \otimes dX_{u_2} = \frac{1}{2}(X_t - X_0) \otimes (X_t - X_0).$$

The mapping $X_{\bullet} \mapsto S_{\bullet}$ is thus continuous in the uniform topology on both sides. The discontinuity we are looking for must lie in the antisymmetric part, which writes

$$A_t = \frac{1}{2} \int_{0<u_1<u_2<t} dX_{u_1} \otimes dX_{u_2} - dX_{u_2} \otimes dX_{u_1}.$$

This tensor has a nice geometric interpretation. Assume that V is a Euclidean space and that $(e_1, \ldots, e_n)$ is an orthonormal basis. Decompose $X = (X^1, \ldots, X^n)$ accordingly. Let $(e_1^*, \ldots, e_n^*)$ be the dual basis. Then, for all i and j distinct integers between 1 and n,

$$(e_i^* \otimes e_j^*)A_t = \int_{0<u_1<u_2<t} dX_{u_1}^i dX_{u_2}^j - dX_{u_1}^i dX_{u_2}^j$$

is the total area enclosed on the plane $\mathrm{Vect}(e_i, e_j)$ by the orthogonal projection of the curve X completed to a closed curve by the straight line from X_t to X_0. This explains the title of this paragraph.

Let us now prove the proposition. In fact, a substantial part of the details are left to the reader as an exercise.

Proof. For each $n \geq 1$, consider the path $X(n) : [0, 2\pi] \longrightarrow \mathbb{R}^2$ defined by

$$X_t(n) = \left(\frac{\cos(n^2 t)}{n}, \frac{\sin(n^2 t)}{n} \right).$$

As n tends to ∞, $X(n)$ tends uniformly to the constant path $X_t = (0,0)$. The symmetric part of the second iterated integral of $X(n)$ tends thus to 0. However, the total area enclosed by $X(n)$ is equal to π, independently of n. Hence, the antisymmetric part of the second iterated integral of $X(n)$ at $t = 1$ does not tend to 0.

We leave it as an exercise for the reader to prove that the sequence $(X(n))_{n \geq 1}$ converges to the constant path at $(0,0)$ in the p-variation norm for every $p > 2$. He is however warned that the convergence does not hold for $p = 2$ and is invited to imagine another sequence of paths, which converges in two-variation to a constant path without the area it encloses tending to 0. $\square$

2

The Signature of a Path

In this chapter, we introduce the second main ingredient of the theory of rough paths, the signature of a path, i.e. the sequence of its definite iterated integrals. This sequence appears naturally when one solves a linear controlled differential equation by iteration. It is also related to some extent to the definition of Lip functions. We will prove in the linear setting that the iterated integrals of a path encode in a very accurate way the information which is necessary to determine the response of a system driven by this path. We will then show that the sequence of all the iterated integrals of a path, which we call its signature, is an extremely interesting object, which is involved in a lot of very rich algebraic structures.

2.1 Iterated Integrals and Linear Equations

Let us consider the following artificial example of a problem described by (1.1). Draw a path $X : [0, \ell] \longrightarrow \mathbb{R}^2$ on a table. Let us assume that this path has finite one-variation and is parametrised by its arc-length, so that, for all $t \in [0, \ell]$, $\|X\|_{1,[0,t]} = t$. Now take a ball of radius R and roll it along X. The general question we want to address is the following: how much about the path X do we need to know in order to be able to predict with a given accuracy the effect of this rolling on the ball?

Let us formulate this question more precisely. At each time $t \in [0, \ell]$, the global motion of the ball since time $t = 0$ is the composition of the translation by $X_t - X_0$ and a certain rotation of $\mathbb{R}^3$ which we represent by a 3×3 orthogonal matrix Φ_t. Now our question is: what information about X should we store in a computer if we want it to be able to compute the matrix Φ_ℓ with a given precision, say with an error less than 2^{-32} on each coefficient?

At this point, the answer is obviously that we should store the matrix Φ_ℓ itself! But what if we do not know the radius R of the ball in advance? Well, we might be in trouble, because if we roll a ball with a very small radius along X, its motion will be very sensitive to a lot of the fine structure of X and as

R tends to 0, more and more information about X will be needed to predict Φ_ℓ with the error we are allowing ourselves. Hence we need at least a lower bound on R: let us assume that $R \geq 1$. Then we claim that there is a good answer to our question and that it involves the iterated integrals of X.

Let us determine by which equations $X_\bullet$ and $\Phi_\bullet$ are related. For this, let us identify the table with the plane $\mathbb{R}^2$ with coordinates (x^1, x^2). Then, if we roll the ball infinitesimally δx^1 in the direction of x^1 (resp., δx^2 in the direction of x^2), it undergoes an infinitesimal rotation $I + \frac{1}{R} A_1 \delta x^1 + O((\delta x^1)^2)$ (resp., $I + \frac{1}{R} A_2 \delta x^2 + O((\delta x^2)^2))$, where I is the identity matrix and

$$A_1 = \begin{pmatrix} 0 & 0 & 1 \\ 0 & 0 & 0 \\ -1 & 0 & 0 \end{pmatrix}, \quad A_2 = \begin{pmatrix} 0 & 0 & 0 \\ 0 & 0 & 1 \\ 0 & -1 & 0 \end{pmatrix}.$$

Hence, if X decomposes as (X^1, X^2) in the coordinates (x^1, x^2), then Φ is the solution to the equation

$$d\Phi_t = \frac{1}{R} A_1 \Phi_t dX_t^1 + \frac{1}{R} A_2 \Phi_t dX_t^2, \quad \Phi_0 = I. \tag{2.1}$$

In the general language of Sect. 1.1.1, we have a control $X : [0, \ell] \longrightarrow \mathbb{R}^2$, a vector field $f : \mathbb{M}_3(\mathbb{R}) \longrightarrow \mathbf{L}(\mathbb{R}^2, \mathbb{M}_3(\mathbb{R}))$ and a response $\Phi : [0, \ell] \longrightarrow \mathbb{M}_3(\mathbb{R})$. Here,

$$f(\Phi)(x^1, x^2) = \frac{1}{R}(A_1 x^1 + A_2 x^2)\Phi.$$

This mapping f is very special in that it is *linear*. Hence, we say that (2.1) is a linear equation. Indeed, linear combinations of solutions, with various initial conditions, are still solutions. Let us give a more formal definition.

Definition 2.1. *Let W and V be two Banach spaces. Let $B : V \longrightarrow \mathbf{L}(W)$ be a bounded linear map. Let $X : J \longrightarrow V$ be a continuous path. Then the equations*

$$dY_t = BY_t \, dX_t, \ Y_0 \in W \tag{2.2}$$

$$d\Phi_t = B\Phi_t \, dX_t, \ \Phi_0 \in \mathbf{L}(W) \tag{2.3}$$

are called linear equations. *In the first one, $BY_t \, dX_t$ means $[B(dX_t)](Y_t)$ and in the second one, $B\Phi_t \, dX_t$ means $B(dX_t) \circ \Phi_t$.*

The mapping $t \mapsto \Phi_t$ is called the flow *associated to the linear equation (2.2).*

In our example, $V = \mathbb{R}^2$, $W = \mathbb{R}^3$ and B is defined by $B(x^1, x^2) = \frac{1}{R}(A_1 x^1 + A_2 x^2)$. Our (2.1) is actually the flow of the differential equation corresponding to (2.2), which describes the motion of a single point on the surface of the ball.

Let us run the standard iterative procedure to solve (2.1). This produces a sequence $(\Phi_\bullet^n)_{n \geq 0}$ of paths from $[0, \ell]$ to $\mathbb{M}_3(\mathbb{R})$ as follows. Set $\Phi_t^0 = I$ for all $t \in [0, \ell]$. Then, set

$$\Phi_t^1 = I + \int_0^t B\Phi_u^0 \, dX_u = I + \int_0^t B(dX_u).$$

Let us do this again and set

$$\Phi_t^2 = I + \int_0^t B\Phi_u^1 \, dX_u = I + \int_0^t B(dX_u) + \int_{0<u_1<u_2<t} B(dX_{u_2})B(dX_{u_1}).$$

For all $x, x' \in \mathbb{R}^2$, set $B^{\otimes 2}(x \otimes x') = B(x')B(x)$. Then we have

$$\Phi_t^2 = I + B \int_{0<u_1<t} dX_{u_1} + B^{\otimes 2} \int_{0<u_1<u_2<t} dX_{u_1} \otimes dX_{u_2}.$$

By iterating this process, we get for all $n \geq 1$ the relation

$$\Phi_t^n = I + \sum_{k=1}^{n} B^{\otimes k} \int_{0<u_1<\ldots<u_k<t} dX_{u_1} \otimes \ldots \otimes dX_{u_k},$$

where we have set $B^{\otimes k}(x_1 \otimes \ldots \otimes x_k) = B(x_k)\ldots B(x_1)$. Proposition 2.2 ensures that the iteration procedure converges.

Proposition 2.2. *Let $X : [0,T] \longrightarrow V$ be a path with finite one-variation. Then, for each $k \geq 1$, one has*

$$\left| \int_{0<u_1<\ldots<u_k<T} dX_{u_1} \otimes \ldots \otimes dX_{u_k} \right| \leq \frac{\|X\|_{1,[0,T]}^k}{k!}.$$

Proof. By reparametrising X, we may assume that $\|X\|_{1,[0,t]} = t$ for all $t \in [0,T]$. Then X is in particular Lipschitz continuous hence almost everywhere differentiable with $|\dot{X}_t| = 1$. Then,

$$\left| \int_{0<u_1<\ldots<u_k<T} dX_{u_1} \otimes \ldots \otimes dX_{u_k} \right| = \left| \int_{0<u_1<\ldots<u_k<T} \dot{X}_{u_1} \otimes \ldots \otimes \dot{X}_{u_k} \, du_1 \ldots du_k \right|$$

$$\leq \int_{0<u_1<\ldots<u_k<T} du_1 \ldots du_k$$

$$= \frac{T^k}{k!}.$$

Since $T = \|X\|_{1,[0,T]}$, this proves the result. $\qquad\square$

We have now a rapidly convergent expansion of the solution Φ to (2.1): for all $t \in [0,\ell]$,

$$\Phi_t = I + \sum_{k=1}^{\infty} B^{\otimes k} \int_{0<u_1<\ldots<u_k<t} dX_{u_1} \otimes \ldots \otimes dX_{u_k}.$$

Moreover, we have the following estimate of the error. For all $n \geq 1$,

$$|\Phi_\ell - \Phi_\ell^n| \leq |I| \sum_{k=n+1}^{\infty} \frac{(\|B\|\|X\|_{1,[0,\ell]})^k}{k!}.$$

This allows us to answer our original question. If $\ell \leq 1$ and since we are assuming that $R \geq 1$, the last equation tells us that, if we keep in the expansion of Φ_ℓ only the terms up to $n = 13$, then we are making an error which is smaller than 2^{-32}. Thus, if we set B_1 equal to the value of B when $R = 1$, then we need only to store the 3×3 matrices

$$D_k = B_1^{\otimes k} \int_{0 < u_1 < \ldots < u_k < \ell} dX_{u_1} \otimes \ldots \otimes dX_{u_k}, \quad k = 1, \ldots, 13$$

with a precision of order 2^{-32} in order to determine Φ_ℓ with the same precision for all $R \geq 1$ by the formula

$$\Phi_\ell \approx I + \sum_{k=1}^{13} \frac{D_k}{R^k}.$$

In fact, we have even got a much stronger result, because the estimate of the error that we make by truncating the expansion of Φ_ℓ depends on the particular linear equation that we are solving only through $\|B\|$. Thus, if we store the values of the first 13 iterated integrals of X with enough precision, then we will be able to compute accurately the response of *any* linear equation driven by X, provided the norm of the one-form B is not too large. Of course, with numerical applications in mind, one must not forget that the successive iterated integrals belong to vector spaces whose dimension grows exponentially fast. Nevertheless, this shows that the iterated integrals of X give a 'top down' analysis of X which enables one to extract the crucial features of X affecting the solution of any linear differential equation driven by X. It will appear in the next chapters that one can drop the word 'linear' in the last sentence.

Remark 2.3. The Young integral allows us to define the iterated integrals of any path with finite p-variation for $p < 2$. Then it is natural to ask whether the arguments above could be applied to linear equations driven by such paths. The answer is positive, but the crucial argument, which is the factorial decay, is more difficult to establish.

2.2 The Signature of a Path

The example of the linear equations shows the importance of the iterated integrals of a path as isolated objects. We are now going to explain how a huge amount of structure arises when one puts together all the iterated integrals of a path into a single object. Let us first describe in which space this single object lives. Throughout this section, we fix a Banach space E in which our paths take their values. Whenever we feel it more comfortable, we will make the assumption that E is finite dimensional, and you are welcome to do it from now on if you prefer. However, this assumption is nowhere really needed in what follows.

2.2.1 Formal Series of Tensors

In Remark 1.24, we have explained that the successive tensor powers of E can be identified with the spaces of homogeneous non-commuting polynomials of successive degrees in as many variables as the elements of a basis of E. We are now going to consider the spaces of non-commuting power series in these variables. We use the convention $E^{\otimes 0} = \mathbb{R}$.

Definition 2.4. *The space of formal series of tensors of E, denoted by $T((E))$, is defined to be the following space of sequences:*

$$T((E)) = \{\mathbf{a} = (a_0, a_1, \ldots) \mid \forall n \geq 0, a_n \in E^{\otimes n}\}.$$

It is endowed with two internal operations, an addition and a product, which are defined as follows. Let $\mathbf{a} = (a_0, a_1, \ldots)$ and $\mathbf{b} = (b_0, b_1, \ldots)$ be two elements of $T((E))$. Then

$$\mathbf{a} + \mathbf{b} = (a_0 + b_0, a_1 + b_1, \ldots)$$

and

$$\mathbf{a} \otimes \mathbf{b} = (c_0, c_1, \ldots),$$

where for each $n \geq 0$,

$$c_n = \sum_{k=0}^{n} a_k \otimes b_{n-k}.$$

The product $\mathbf{a} \otimes \mathbf{b}$ is also denoted by $\mathbf{ab}$.

The space $T((E))$ endowed with these two operations and the natural action of $\mathbb{R}$ by $\lambda \mathbf{a} = (\lambda a_0, \lambda a_1, \ldots)$ is a real non-commutative unital algebra, with unit $\mathbf{1} = (1, 0, 0, \ldots)$.

An element $\mathbf{a} = (a_0, a_1, \ldots)$ of $T((E))$ is invertible if and only if $a_0 \neq 0$. Its inverse is then given by the series

$$\mathbf{a}^{-1} = \frac{1}{a_0} \sum_{n \geq 0} \left(1 - \frac{\mathbf{a}}{a_0}\right)^n,$$

which is well defined because, for each given degree, only finitely many terms of the sum produce non-zero tensors of this degree. In particular, the subset

$$\tilde{T}((E)) = \{\mathbf{a} \in T((E)) \mid a_0 = 1\}$$

is a group. The signature of a path will be defined as an element of this group.

It is often important to look only at finitely many terms of an element of $T((E))$. For each $n \geq 0$, the space $B_n = \{\mathbf{a} = (a_0, a_1, \ldots) \mid a_0 = \ldots = a_n = 0\}$ of formal series with no monomials of degree less or equal to n is an ideal of $T((E))$.

Definition 2.5. *Let $n \geq 1$ be an integer. The truncated tensor algebra of order n of E is defined as the quotient algebra*

$$T^{(n)}(E) = T((E))/B_n.$$

The canonical homomorphism $T((E)) \longrightarrow T^{(n)}(E)$ is denoted by π_n.

In fact, $T^{(n)}(E)$ is canonically isomorphic to $\bigoplus_{k=0}^{n} E^{\otimes k}$ equipped with the product

$$(a_0, \ldots, a_n)(b_0, \ldots, b_n) = (c_0, \ldots, c_n),$$

where, for all $k \in \{0, \ldots, n\}$, $c_k = a_0 \otimes b_k + a_1 \otimes b_{k-1} + \ldots + a_k \otimes b_0$. The homomorphism π_n consists then simply in forgetting the terms of degree greater than n. However, when one thinks of $T^{(n)}(E)$ in that way, one must keep in mind that $T^{(n)}(E)$ is embedded in $T((E))$ as a linear subspace, but not as a sub-algebra.

2.2.2 The Signature of a Path

Definition 2.6. *Let J denote a compact interval. Let $X : J \longrightarrow E$ be a continuous path with finite p-variation for some $p < 2$. The signature of X is the element $\mathbf{X}$ of $T((E))$ defined as follows*

$$\mathbf{X}_J = (1, X_J^1, X_J^2, \ldots),$$

where, for each $n \geq 1$, $X_J^n = \displaystyle\int_{\substack{u_1 < \ldots < u_n \\ u_1, \ldots, u_n \in J}} dX_{u_1} \otimes \ldots \otimes dX_{u_n}$. The signature of X is also denoted by $S(X)$.

Remark 2.7. Since we are persistently claiming that the signature encodes so much information about the path, the reader may wonder if it determines the path completely or not. In fact, when $p > 1$, this is still an open question. We will come back to this problem later in this chapter and give the answer for $p = 1$.

The signature is a mapping of a set of paths, modulo translations and reparametrisations, into an algebra. It is time to notice that paths, modulo translations and reparametrisations have indeed a multiplication!

Definition 2.8. *Let $X : [0, s] \longrightarrow E$ and $Y : [s, t] \longrightarrow E$ be two continuous paths. Their concatenation is the path $X * Y : [0, t] \longrightarrow E$ defined by*

$$(X * Y)_u = \begin{cases} X_u & \text{if } u \in [0, s] \\ X_s + Y_u - Y_s & \text{if } u \in [s, t]. \end{cases}$$

The following fundamental theorem of Chen [1] asserts that the signature is a homomorphism.

Theorem 2.9 (Chen). *Let $X : [0, s] \longrightarrow E$ and $Y : [s, t] \longrightarrow E$ be two continuous paths with finite one-variation. Then*

$$S(X * Y) = S(X) \otimes S(Y).$$

Proof. Set $Z = X * Y$ and $S(Z) = (1, Z_{0,t}^1, Z_{0,t}^2, \ldots)$. Let us choose $n \geq 1$ and compute $Z_{0,t}^n$. It is equal to

$$Z_{0,t}^n = \int \cdots \int_{0 < u_1 < \ldots < u_n < t} dZ_{u_1} \otimes \ldots \otimes dZ_{u_n}$$

$$= \sum_{k=0}^n \int \cdots \int_{0 < u_1 < \ldots < u_k < s < u_{k+1} < \ldots < u_n < t} dZ_{u_1} \otimes \ldots \otimes dZ_{u_n}.$$

By Fubini's theorem, this is equal to

$$Z_{0,t}^n = \sum_{k=0}^n \int \cdots \int_{0 < u_1 < \ldots < u_k < s} dX_{u_1} \otimes \ldots \otimes dX_{u_k} \otimes \int \cdots \int_{s < u_{k+1} < \ldots < u_n < t} dY_{u_{k+1}} \otimes \ldots \otimes dY_{u_n}$$

$$= \sum_{k=0}^n X_{0,s}^k \otimes Y_{s,t}^{n-k}.$$

Hence, $S(Z) = S(X) \otimes S(Y)$. $\qquad\square$

What about paths with finite p-variation for some $p \in (1, 2)$? For them, we cannot use Fubini's theorem and the proof of Chen's theorem does not apply. However, Lemma 2.10 will allow us to extend Chen's theorem.

Lemma 2.10. *Let $X : [0, T] \longrightarrow E$ be a path with finite p-variation for some $p < 2$. Define $f : T^{(n)}(E) \longrightarrow \mathbf{L}(E, T^{(n)}(E))$ by*

$$f(a_0, a_1, \ldots, a_n)x = (0, a_0 \otimes x, a_1 \otimes x, \ldots, a_{n-1} \otimes x).$$

Then the unique solution to the differential equation

$$dS_t = f(S_t)\, dX_t, \quad S_0 = (1, 0, \ldots, 0) \tag{2.4}$$

is the path $S : [0, T] \longrightarrow T^{(n)}(E)$ defined for all $t \in [0, T]$ by

$$S_t = \pi_n(S(X_{|[0,t]})) = (1, X_{0,t}^1, \ldots, X_{0,t}^n).$$

Proof. The existence and uniqueness of the solution to (2.4) are a consequence of Theorem 1.28. We invite the reader to check that $S_\bullet$ defined by the last equality is indeed the solution. $\qquad\square$

However easy to prove, this lemma is extremely important and useful. The element $\pi_n(S(X))$ is called the *truncated signature* of X of order n.

Corollary 2.11. *Let J be a compact interval. For each $p \in [1,2)$ and each integer $n \geq 0$, the truncated signature is a continuous mapping*

$$\pi_n \circ S : \mathcal{V}^p(J, E) \longrightarrow T^{(n)}(E).$$

Proof. By Lemma 2.10, the mapping $\pi_n \circ S$ is the Itô map for a polynomial vector field. Such a vector field, restricted to a bounded neighbourhood of the range of any path, is Lip(γ) for every $\gamma > 0$. Hence, by Theorem 1.28, the mapping $\pi_n \circ S$ is continuous in p-variation norm. $\square$

Remark 2.12. If we use bold letters for the elements of $T^{(n)}(E)$ and recall that E is embedded in $T^{(n)}(E)$, then the vector field f in (2.4) can be rewritten as $f(\mathbf{a})x = \mathbf{a} \otimes x$. Since the truncated signature of order n is the solution of (2.4) for each $n \geq 0$, we can formally write the following differential equation for the full signature of $X_{|[0,t]}$ as a function from $[0,T]$ to $T((E))$:

$$d\mathbf{X}_{0,t} = \mathbf{X}_{0,t} \otimes dX_t, \quad \mathbf{X}_{0,0} = \mathbf{1}.$$

The signature of a path can be described as its full non-commutative exponential.

Corollary 2.13. *Chen's theorem (Theorem 2.9) holds for paths with finite p-variation for $p < 2$.*

Proof. Let $p < 2$ be given and choose p' such that $p < p' < 2$. Let $X : [0, s] \longrightarrow E$ and $Y : [s, t] \longrightarrow E$ be two continuous paths with finite p-variation. Let $(X(m))_{m \geq 0}$ and $(Y(m))_{m \geq 0}$ be two sequence of paths with finite one-variation which converge, respectively, to X and Y in p'-variation. Such sequences exist by Proposition 1.14. By Theorem 2.9, for each $m \geq 0$, $S(X(m) * Y(m)) = S(X(m)) \otimes S(Y(m))$. In particular, for each $n \geq 0$, since π_n is an algebra homomorphism, we have $(\pi_n \circ S)(X(m) * Y(m)) = (\pi_n \circ S)(X(m)) \otimes (\pi_n \circ S)(Y(m))$.

As m tends to ∞, both sides of this equality converge and we find

$$\pi_n(S(X * Y)) = \pi_n(S(X)) \otimes \pi_n(S(Y)) = \pi_n(S(X) \otimes S(Y)).$$

Since $S(X * Y)$ and $S(X) \otimes S(Y)$ agree up to degree n for every $n \geq 0$, they are equal. $\square$

The range of the mapping $S : \mathcal{V}^p(J, E) \longrightarrow T((E))$ is an important object, and not an easy one to describe. Chen's theorem asserts that it is closed under multiplication. Now let X be a path with finite p-variation. Then, since the term of degree 0 in $S(X)$ is 1, we know that $S(X)$ is invertible. However, we still need to prove that $S(X)^{-1}$ is the signature of a path.

Proposition 2.14. *Let* $X : [0,T] \longrightarrow E$ *be a path with finite p-variation for* $p < 2$. *Let* $\overleftarrow{X}$ *be the path* X *run backwards, i.e. the path defined by* $\overleftarrow{X}_t = X_{T-t}$, $t \in [0,T]$. *Then*

$$S(\overleftarrow{X}) = S(X)^{-1}.$$

In particular, the range of $S : \mathcal{V}^p([0,T],E) \longrightarrow T((E))$ *is a group.*

Proof. It would be rather painful to prove this result by computing directly $S(X) \otimes S(\overleftarrow{X})$ from the definition. Instead, let us use again Lemma 2.10. For this, set $N = X * \overleftarrow{X} : [0,2T] \longrightarrow E$. Let V be a Banach space. Let $f : V \longrightarrow \mathbf{L}(E,V)$ be a $\mathrm{Lip}(\gamma)$ one-form with $\gamma > p$. Then it is equivalent for a path $Y : [0,T] \longrightarrow V$ to satisfy

$$\forall t \in [0,T], \quad dY_t = f(Y_t)\,dX_t, \quad Y_0 = \xi, \ Y_T = \eta,$$

or to satisfy

$$\forall t \in [0,T], \quad d\overleftarrow{Y}_t = f(\overleftarrow{Y}_t)\,d\overleftarrow{X}_t, \quad \overleftarrow{Y}_0 = \eta, \ \overleftarrow{Y}_T = \xi.$$

Hence, whatever the one-form f is, the solution to the equation

$$dY_t = f(Y_t)\,dN_t, \quad Y_0 = \xi$$

satisfies $Y_{2T} = \xi$. In particular, after taking for f the one-form of (2.4), Lemma 2.10 implies that, for all $n \geq 0$, $\pi_n(S(N)) = \pi_n(\mathbf{1})$. Hence, $S(X) \otimes S(\overleftarrow{X}) = S(N) = \mathbf{1}$. $\qquad\square$

In order to better understand the range of the signature, we are going to study an algebra of functions on this subset of $T((E))$.

2.2.3 Functions on the Range of the Signature

Whenever one does analysis, one works hard to find a core of real functions on the space one is working on, and ideally this core should be an algebra. When probabilists study spaces of paths, they tend to use cylinder functions. We now present a different algebra of functions, with many excellent properties.

These functions are induced by *linear forms* on $T((E))$, which we start by describing. For this, assume that E has finite dimension and let $(e_1, \ldots, e_d)$ be a basis of E. Let $(e_1^*, \ldots, e_d^*)$ be the dual basis of E^*. Let $n \geq 1$ be an integer. The elements $(e_I = e_{i_1} \otimes \ldots \otimes e_{i_n})_{I=(i_1,\ldots,i_n) \in \{1,\ldots,d\}^n}$ form a basis of $E^{\otimes n}$. Then $(E^*)^{\otimes n}$ can be canonically identified with $(E^{\otimes n})^*$ by identifying the basis $(e_I^* = e_{i_1}^* \otimes \ldots \otimes e_{i_n}^*)_{I=(i_1,\ldots,i_n) \in \{1,\ldots,d\}^n}$ of $(E^*)^{\otimes n}$ with the dual basis of $(E^{\otimes n})^*$. In other words,

$$\langle e_{i_1}^* \otimes \ldots \otimes e_{i_n}^*, e_{j_1} \otimes \ldots \otimes e_{j_n} \rangle = \delta_{i_1,j_1} \ldots \delta_{i_n,j_n}.$$

In fact, the linear action of $(E^*)^{\otimes n}$ on $E^{\otimes n}$ extends naturally to a linear mapping $(E^*)^{\otimes n} \longrightarrow T((E))^*$, defined by

$$e_I^*(\mathbf{a}) = e_I^*(a_n).$$

If we think of $\mathbf{a}$ as a non-commuting power series in the letters $e_1, \ldots, e_d$, then $e^*_{(i_1, \ldots, i_n)}$ picks up the coefficient of the monomial $e_{i_1} \ldots e_{i_n}$.

By letting n vary between 0 and ∞, we get a linear mapping

$$T(E^*) = \bigoplus_{n=0}^{\infty} (E^*)^{\otimes n} \longrightarrow T((E))^*.$$

The elements of $T(E^*)$ are the sequences of elements of the successive tensor powers of E^* whose elements are all but a finite number equal to 0. Hence, the linear forms e^*_I, as I span the set of finite words in the letters $1, \ldots, d$, form a basis of $T(E^*)$.

Let us choose a word $I = (i_1, \ldots, i_n)$ and a path $X : [0, T] \longrightarrow E$ with finite p-variation for some $p < 2$. Then, we may form the real number

$$\varphi_I(X) = e^*_I(S(X)) = \int \cdots \int_{0 < u_1 < \ldots < u_n < T} e^*_{i_1}(dX_{u_1}) \ldots e^*_{i_n}(dX_{u_n}).$$

In other words, $\varphi_I = e^*_I \circ S$ is a real-valued function on $\mathcal{V}^p([0, T], E)$. By linearity, every element $\mathbf{e}^* \in T(E^*)$ determines a real-valued function $\varphi_{\mathbf{e}^*} = \mathbf{e}^* \circ S$ on $\mathcal{V}^p([0, T], E)$.

Now take two elements of $T(E^*)$, say $\mathbf{e}^*$ and $\mathbf{f}^*$. Their pointwise product as real-valued functions is a quadratic form on $T((E))$. Then it is well understood, but still remarkable, that there exists a unique third element of $T(E^*)$, denoted by $\mathbf{e}^* \sqcup \mathbf{f}^*$, such that $\mathbf{e}^* \mathbf{f}^*$ and $\mathbf{e}^* \sqcup \mathbf{f}^*$ coincide on the range of S. We do not prove the uniqueness of this third form for the moment.

Theorem 2.15. *The linear forms on $T((E))$ induced by $T(E^*)$, when restricted to the range $S(\mathcal{V}^p([0, T], E))$ of the signature, form an algebra of real-valued functions.*

Proof. By linearity, it is sufficient to prove the result when $\mathbf{e}^* = e^*_I$ and $\mathbf{f}^* = e^*_J$ for some finite words $I = (i_1, \ldots, i_r)$ and $J = (j_1, \ldots, j_s)$. Once two such words are fixed, set $(k_1, \ldots, k_{r+s}) = (i_1, \ldots, i_r, j_1, \ldots, j_s)$. Now we say that a permutation $\sigma \in \mathfrak{S}_{r+s}$ is a *shuffle* of $\{1, \ldots, r\}$ and $\{r+1, \ldots, r+s\}$ if $\sigma(1) < \ldots < \sigma(r)$ and $\sigma(r+1) < \ldots < \sigma(s)$. We write $\sigma \in \text{Shuffles}(r, s)$.

Let $X : [0, T] \longrightarrow E$ be a path with bounded variation. Then $\varphi_I(X)\varphi_J(X)$ is equal to

$$\int \cdots \int_{0 < u_1 < \ldots < u_r < T} e^*_{i_1}(dX_{u_1}) \ldots e^*_{i_r}(dX_{u_r}) \int \cdots \int_{0 < v_1 < \ldots < v_s < T} e^*_{j_1}(dX_{v_1}) \ldots e^*_{j_s}(dX_{v_s})$$

$$= \sum_{\sigma \in \text{Shuffles}(r,s)} \int \cdots \int_{0 < w_1 < \ldots < w_{r+s} < T} e^*_{k_{\sigma^{-1}(1)}}(dX_{w_1}) \ldots e^*_{k_{\sigma^{-1}(r+s)}}(dX_{w_{r+s}}).$$

We have proved that the linear form

$$e_I^* \sqcup e_J^* = \sum_{\sigma \in \text{Shuffles}(r,s)} e_{(k_{\sigma^{-1}(1)}, \ldots, k_{\sigma^{-1}(r+s)})}^* \tag{2.5}$$

is equal to $e_I^* e_J^*$ on the subset $S(\mathcal{V}^1([0,T]), E)$ of $T((E))$. Observe that we have used Fubini's theorem, the reason for which we insisted that X has bounded variation. If X has only finite p-variation for some $p \in (1,2)$, then one uses the fact that, for each $\mathbf{e}^* \in T(E^*)$, the equality $\mathbf{e}^* \circ \pi_n = \mathbf{e}^*$ holds for n large enough, and an approximation argument similar to that used in the proof of Corollary 2.13. We leave the details to the reader. $\square$

The linear form $e_I^* \sqcup e_J^*$ defined by (2.5) is called the *shuffle product* of e_I^* and e_J^*. By bilinearity, this product endows $T(E^*)$ with a structure of algebra, known as the shuffle algebra (*algèbre de battage* in French).

Corollary 2.16. *If the signatures of a finite collection of paths with finite p-variation for some $p < 2$ are pairwise distinct, then they are linearly independent.*

Proof. Let $X_1, \ldots, X_n$ be n paths with bounded variation and pairwise distinct signatures. On the finite set $\{S(X_1), \ldots, S(X_n)\}$, the elements of $T(E^*)$ induce an algebra of real-valued functions. This algebra contains the constants, because the first term of every signature is 1. Moreover, it separates the points. Indeed, if two elements $\mathbf{s_1}$ and $\mathbf{s_2}$ of $T((E))$ are distinct, there exists $n \geq 0$ such that $\pi_n(\mathbf{s_1}) \neq \pi_n(\mathbf{s_2})$. Since $T^{(n)}(E)$ is finite dimensional, there exists a linear form $\sigma \in T^{(n)}(E)^*$ which separates $\pi_n(\mathbf{s_1})$ and $\pi_n(\mathbf{s_1})$. Now, $T^{(n)}(E)^* = T^{(n)}(E^*) \subset T(E^*)$, so that σ can be identified with an element of $T(E^*)$ which, moreover, satisfies $\sigma \circ \pi_n = \sigma$. Hence, $\sigma(\mathbf{s_1}) \neq \sigma(\mathbf{s_2})$.

This implies that, for each $i \in \{1, \ldots, n\}$, there exists $\mathbf{e}^* \in T((E))^*$ such that $\mathbf{e}^*(S(X_k)) = 1$ if $k = i$ and 0 otherwise. This terminates the proof. $\square$

We have defined a product on $T(E^*)$ and proved that, for all $\mathbf{e}^*, \mathbf{f}^* \in T(E^*)$ and for all $\mathbf{a} \in S(\mathcal{V}^p([0,T], E))$

$$\mathbf{e}^*(\mathbf{a})\mathbf{f}^*(\mathbf{a}) = (\mathbf{e}^* \sqcup \mathbf{f}^*)(\mathbf{a}). \tag{2.6}$$

It is tempting to ask whether this property characterises the range of the signature. Lemma 2.17 is left as an exercise for the reader. Recall that $\widetilde{T}((E))$ denotes the set of formal series whose constant term is 1.

Lemma 2.17. *The subset of $\widetilde{T}((E))$ where the identity (2.6) holds for all $\mathbf{e}^*, \mathbf{f}^* \in T(E^*)$ is a group.*

Definition 2.18. *An element $\mathbf{a} \in \widetilde{T}((E))$ is said to be group-like if the evaluation mapping $\mathrm{ev}_\mathbf{a} : T(E^*) \longrightarrow \mathbb{R}$ defined by $\mathrm{ev}_\mathbf{a}(\mathbf{e}^*) = \mathbf{e}^*(\mathbf{a})$ is a morphism of algebras when $T(E^*)$ is endowed with the shuffle product. The set of group-like elements is denoted by $G^{(*)}$.*

We are going to state an equivalent characterisation of group-like elements, which explains their name and will allow us to prove that the range of the signature is a proper subgroup of $G^{(*)}$, indeed a rather small one. On the other hand, it will turn out that, for each $n \geq 0$, the range of the truncated signature $\pi_n \circ S$ coincides with the group $G^{(n)} = \pi_n(G^{(*)})$, which is the free nilpotent group of order n.

2.2.4 Lie Elements, Logarithm and Exponential

In order to define the exponential of a formal series, we make the assumption that the tensor powers of E are endowed with admissible norms in the sense of Definition 1.25. For example, if we think of tensors as non-commuting polynomials, then the norms of the supremum of the coefficients are admissible, as well as the ℓ^1 norms. Under this assumption, the following property is easily checked.

Lemma 2.19. *Let* $\mathbf{a}$ *be an element of* $T((E))$. *Then the series*

$$\sum_{n \geq 0} \frac{\mathbf{a}^n}{n!}$$

is convergent. In other words, for each $k \geq 0$, *the term of degree* k *of* $\sum_{n=0}^{N} \frac{\mathbf{a}^n}{n!}$ *has a limit in* $E^{\otimes k}$ *as* N *tends to* ∞.

Definition 2.20. *Let* $\mathbf{a}$ *be an element of* $T((E))$. *Then* $\exp(\mathbf{a})$ *is the element of* $\widetilde{T}((E))$ *defined by*

$$\exp \mathbf{a} = \sum_{n=0}^{\infty} \frac{\mathbf{a}^n}{n!}.$$

If $a_0 > 0$, *then* $\log(\mathbf{a})$ *is the element of* $T((E))$ *defined by*

$$\log \mathbf{a} = \log(a_0) + \sum_{n \geq 1} \frac{(-1)^n}{n} \left(1 - \frac{\mathbf{a}}{a_0} \right)^n.$$

The series defining $\log \mathbf{a}$ is locally finite, in that it produces only finitely many terms of each degree. It is thus purely algebraic and does not depend on the norms on the tensor powers of E.

Recall that B_1 is the subset of $T((E))$ which consists in the series whose constant term is 0. Lemma 2.21 is elementary and we leave it to the reader.

Lemma 2.21. *The mappings* $\exp : B_1 \longrightarrow \widetilde{T}((E))$ *and* $\log : \widetilde{T}((E)) \longrightarrow B_1$ *are one-to-one and they are each other's inverse.*

It turns out that the subset $\log(G^{(*)})$ of $T((E))$ can be described quite concretely. For this, observe that the algebra $T((E))$, as any associative algebra, carries a Lie bracket $[\cdot, \cdot]$ defined by the relation

$$[\mathbf{a}, \mathbf{b}] = \mathbf{a} \otimes \mathbf{b} - \mathbf{b} \otimes \mathbf{a}.$$

If F_1 and F_2 are two linear subspaces of $T((E))$, let us denote by $[F_1, F_2]$ the linear span of all the elements of the form $[\mathbf{a}, \mathbf{b}]$, where $\mathbf{a} \in F_1$ and $\mathbf{b} \in F_2$.

Consider the sequence $(L_n)_{n \geq 0}$ of subspaces of $T((E))$ defined recursively as follows:

$$L_0 = 0, \quad L_1 = E, \quad L_2 = [E, L_1] = [E, E], \quad L_3 = [E, L_2] = [E, [E, E]], \ldots$$

and, for all $n \geq 1$, $L_{n+1} = [E, L_n]$. Observe that, for each n, L_n is a linear subspace of $E^{\otimes n}$. It is called the space of *homogeneous Lie polynomials* of degree n.

Definition 2.22. *The space of Lie formal series over* E, *denoted as* $\mathcal{L}((E))$, *is defined as the following subspace of* $T((E))$:

$$\mathcal{L}((E)) = \{\mathbf{l} = (l_0, l_1, \ldots) \mid \forall n \geq 0, l_n \in L_n\}.$$

Theorem 2.23 is a classical result of the theory of free Lie algebras. See [75, Theorem 3.2] for a proof.

Theorem 2.23. *An element of* $\widetilde{T}((E))$ *is group-like if and only if its logarithm is a Lie series. In other words,*

$$G^{(*)} = \exp \mathcal{L}((E)) \quad and \quad \log G^{(*)} = \mathcal{L}((E)).$$

An important consequence of this theorem is that, if X is a path with finite p-variation for some $p < 2$, then $\log S(X)$, the log-signature of X, is a Lie series. This result was first proved by Chen and it implies in particular the Baker–Campbell–Hausdorff theorem. Indeed, let x and y be two linearly independent elements of E. Let X be a straight path from 0 to x, and Y be a straight path from 0 to y. Then $S(X) = \exp x$ and $S(Y) = \exp y$. Now, the fact that $S(X * Y) = S(X) \otimes S(Y) = \exp x \otimes \exp y$ is the exponential of a Lie series is exactly the content of the Baker–Campbell–Hausdorff theorem.

2.2.5 Truncated Signature and Free Nilpotent Groups

For each $n \geq 1$, set $G^{(n)} = \pi_n(G^{(*)})$ and $\mathcal{L}^{(n)}(E) = \pi_n[\mathcal{L}((E))]$. The elements of $\mathcal{L}^{(n)}(E)$ are called *Lie polynomials* of degree n.

Set $\widetilde{T}^{(n)}(E) = \pi_n[\widetilde{T}((E))]$. The same power series which defines $\log : \widetilde{T}((E)) \longrightarrow T((E))$ defines a mapping $\widetilde{T}^{(n)}(E) \longrightarrow T^{(n)}(E)$, which is denoted by $\log^{(n)}$. Moreover, if $\mathbf{a}$ belongs to $\widetilde{T}((E))$, then $\pi_n(\log \mathbf{a})$ depends only on the first n terms of the power series, so that $\pi_n(\log \mathbf{a}) = \log^{(n)}(\pi_n(\mathbf{a}))$.

Lemma 2.24 is not difficult to prove but it is important.

Lemma 2.24. *1. Pick $n \geq 0$ and $\mathbf{a} \in \widetilde{T}^{(n)}(E)$. Then $\mathbf{a}$ belongs to $G^{(n)}$ if and only if $\log^{(n)} \mathbf{a}$ belongs to $\mathcal{L}^{(n)}(E)$.*

2. Let $\mathbf{a}$ be an element of $\widetilde{T}((E))$. Then $\mathbf{a}$ is group-like if and only if, for each $n \geq 0$, $\pi_n(a)$ belongs to $G^{(n)}$.

Proof. 1. Assume that $\mathbf{a} = \pi_n(\mathbf{b})$ for some $\mathbf{b} \in G^{(*)}$. Then $\log \mathbf{a} = \log \pi_n \mathbf{b} = \pi_n \log \mathbf{b}$ belongs to $\mathcal{L}^{(n)}(E)$.

Assume now that $\mathbf{l} = (0, l_1, \ldots, l_n) = \log \mathbf{a}$ is a Lie polynomial. Then the fact that $\mathbf{a} = \pi_n(\exp(0, l_1, \ldots, l_n, 0, \ldots))$ shows that $\mathbf{a}$ belongs to $G^{(n)}$.

2. It is equivalent to say that $\mathbf{a}$ is group-like, or to say that, for each $n \geq 0$, $\pi_n(\log \mathbf{a})$ is Lie polynomial, or to say that, for each $n \geq 0$, $\log(\pi_n(\mathbf{a}))$ is a Lie polynomial, or finally to say that, for each $n \geq 0$, $\pi_n(\mathbf{a})$ belongs to $G^{(n)}$. $\square$

The power series of the exponential makes perfect sense on $T^{(n)}(E)$ and it defines there a new mapping, denoted by $\exp^{(n)}$. It is easy to check that $\exp^{(n)} : \pi_n(B_1) \longrightarrow \widetilde{T}^{(n)}(E)$ and $\log^{(n)} : \widetilde{T}^{(n)}(E) \longrightarrow \pi_n(B_1)$ are each other's inverse. If E is finite dimensional, then the following result follows immediately from this observation and the first part of Lemma 2.24.

Proposition 2.25. *The group $G^{(n)}$ is a closed Lie subgroup of $\widetilde{T}^{(n)}(E)$ with Lie algebra $\mathcal{L}^{(n)}(E)$. Moreover, the exponential map of this Lie group is a global diffeomorphism.*

The group $G^{(n)}$ is called the *free nilpotent group of step n over E*. It is globally diffeomorphic to its Lie algebra $E \oplus [E, E] \oplus \ldots \oplus [E, [E, \ldots, [E, E] \ldots]]$ and, in this global chart, the product of two elements is given by the Baker–Campbell–Hausdorff formula.

The definition of this group allows us to give the following interpretation of the truncated signature. Let X be a piecewise differentiable path in E. By Lemma 2.10, the truncated signature S_t of $X_{|[0,t]}$, as a function of t, is the solution to the differential equation

$$\dot{S}_t = S_t \dot{X}_t, \quad S_0 = \pi_n(\mathbf{1}).$$

In other words, the truncated signature of X is nothing but the terminal point of the development in the Lie group $G^{(n)}$ of the path $X_\bullet$, which is a path in the subspace $E \subset \mathcal{L}^{(n)}(E)$ of its Lie algebra.

As in the equation above, we always identify the elements of the Lie algebra of $G^{(n)}$ with left-invariant vector fields on $G^{(n)}$. Now, and with the terminology of sub-Riemannian geometry, let us call *horizontal* path a piecewise differentiable path in $G^{(n)}$ whose derivative belongs at each point to the subspace E of $\mathcal{L}^{(n)}(E)$. If a horizontal path $\mathbf{X}$ starts at $\mathbf{1}$, then its terminal point is the signature of a piecewise differentiable path, namely the path

$$X_\bullet = \int_0^\bullet \pi_1(\mathbf{X}_t^{-1}\dot{\mathbf{X}}_t)\,dt,$$

where we identify $0 \oplus E$ with E.

Theorem 2.26 is a version of the fundamental theorem of sub-Riemannian geometry.

Theorem 2.26 (Chow–Rashevskii). *Every point of $G^{(n)}$ can be joined to $\mathbf{1}$ by a piecewise differentiable horizontal path.*

The horizontal path can actually be chosen to be smooth (see [39]). In the language of iterated integrals, this theorem says the following.

Proposition 2.27. *Every element of $G^{(n)}$ is the truncated signature of a path with bounded variation. In particular, for every $p \in [1, 2)$, $G^{(n)}$ is exactly the range of the mapping*

$$\pi_n \circ S : \mathcal{V}^p([0, T], E) \longrightarrow \widetilde{T}^{(n)}(E).$$

2.2.6 The Signature of Paths with Bounded Variation

In this last section, we go back to the question posed in Remark 2.7: to what extent does the signature of a path determine this path? We state the answer in the case of paths with bounded variation. The details can be found in [43].

On $\mathcal{V}^1([0, T], E)$, we consider the following relation: we say that $X \sim Y$ if $S(X) = S(Y)$. Our goal is to understand this equivalence relation. To start with, we know several reasons why $X \sim Y$ does not imply $X = Y$. The first is that $S(X)$ does not depend on the parametrisation of X. Another more serious reason is that if $X \in \mathcal{V}^1(E)$ is a non-constant path, then, according to Proposition 2.14, $S(X * \overleftarrow{X}) = \mathbf{1}$. Thus, $X * \overleftarrow{X}$ has the same signature as the constant path, but cannot be reparametrised to be constant. There are more complicated situations: if X, Y, Z are non-constant paths, then $S(X * Y * \overleftarrow{Y} * Z * \overleftarrow{Z} * \overleftarrow{X}) = \mathbf{1}$, but this path is not of the form $W * \overleftarrow{W}$ for any path W. However, it can be reduced to a constant path by successively removing pieces of the form $W * \overleftarrow{W}$. In other words, this path looks like a tree: it is *tree-like*. Tree-like paths are those which can be reduced to a constant path by removing possibly infinitesimal pieces of the form $W * \overleftarrow{W}$.

Definition 2.28. *A path $X : [0, T] \longrightarrow E$ is tree-like if there exists a continuous function $h : [0, T] \longrightarrow [0, +\infty)$ such that $h(0) = h(T) = 0$ and such that, for all $s, t \in [0, T]$ with $s \le t$,*

$$\|X_t - X_s\| \le h(s) + h(t) - 2 \inf_{u \in [s,t]} h(u).$$

The function h is called a *height function* for the path X. The notion of tree-like path allows us to give a very nice description of the relation $\sim$. We do not include a proof of the theorem and instead refer the reader to [43].

Theorem 2.29. *1. Let $X : [0,T] \longrightarrow E$ be a path with bounded variation. Then the signature of X is equal to $\mathbf{1}$ if and only if X is tree-like.*

2. Let $Y : [0,T] \longrightarrow E$ be another path with bounded variation. Then $X \sim Y$ if and only if $X * \overleftarrow{Y}$ is tree-like.

3. There exists among all paths with bounded variation and the same signature as X a path with minimal length. This path is unique up to reparametrisation.

3

Rough Paths

In Chap. 2, we have described an algebraic object associated to every path of bounded variation or finite p-variation for some $p < 2$: its signature. The core idea of the theory of rough paths is to consider the signature, rather than the path, as the fundamental object. In this chapter, we present the analytical framework in which this idea becomes effective. We prove the first fundamental theorem of the theory (Theorem 3.7), give the definition of a rough path and discuss some important topological spaces of rough paths. Finally, we take a more detailed look at the Brownian rough paths.

3.1 Multiplicative Functionals

3.1.1 Definition of Multiplicative Functionals

Let V be a Banach space. Let $x : [0, T] \longrightarrow V$ be a continuous path with finite p-variation for some $p < 2$. Recall that $\triangle_T$ denotes the set $\{(s, t) \in [0, T]^2 : 0 \leq s \leq t \leq T\}$. According to Definition 2.6, let us consider, for each $(s, t) \in \triangle_T$, the signature of $x_{|[s,t]}$. It is an element of the tensor algebra $T((V))$, which we denote by $X_{s,t} = S(x_{|[s,t]})$. It is defined by:

$$X_{s,t} = (1, X_{s,t}^1, X_{s,t}^2, \ldots) \in \mathbb{R} \oplus V \oplus (V \otimes V) \oplus \ldots,$$

with

$$X_{s,t}^n = \int \cdots \int_{s \leq u_1 < \ldots < u_n \leq t} dx_{u_1} \otimes \ldots \otimes dx_{u_n}.$$

Chen's theorem (Theorem 2.9) asserts that the map from $\triangle_T$ to $T((V))$ which sends (s, t) to $X_{s,t}$ is *multiplicative*. This means that for all $s, u, t \in [0, T]$, with $s \leq u \leq t$, one has $X_{s,u} \otimes X_{u,t} = X_{s,t}$.

Furthermore, according to Definition 2.5, $\pi_n : T((V)) \longrightarrow T^{(n)}(V)$ is an algebra homomorphism, and hence the mapping from $\triangle_T$ into the nth truncated tensor algebra $T^{(n)}(V)$, which sends (s,t) to the truncated signature $\pi_n(X_{s,t}) = X_{s,t}^{(n)} = (1, X_{s,t}^1, \dots, X_{s,t}^n)$, is also multiplicative.

In what follows, we are going to take as our basic object of study a general function from $\triangle_T$ into a truncated tensor algebra of V which satisfies a multiplicativity property.

Definition 3.1 (Multiplicative Functional). *Let $n \geq 1$ be an integer. Let $X : \triangle_T \longrightarrow T^{(n)}(V)$ be a continuous map. For each $(s,t) \in \triangle_T$, denote by $X_{s,t}$ the image by X of (s,t) and write*

$$X_{s,t} = (X_{s,t}^0, X_{s,t}^1, \dots X_{s,t}^n) \in \mathbb{R} \oplus V \oplus V^{\otimes 2} \oplus \dots \oplus V^{\otimes n}.$$

The function X is called a multiplicative functional of degree n in V if $X_{s,t}^0 = 1$ for all $(s,t) \in \triangle_T$ and

$$X_{s,u} \otimes X_{u,t} = X_{s,t} \quad \forall s, u, t \in [0,T] , \quad s \leq u \leq t. \tag{3.1}$$

The multiplicative property (3.1) is called Chen's identity, in reference to Chen's theorem (Theorem 2.9). In order to get a better understanding of what this property means, let us look first at the special cases $n = 1$ and $n = 2$.

Example 3.2. Consider a functional $X : \triangle_T \longrightarrow T^{(1)}(V) = \mathbb{R} \oplus V$ which is multiplicative. Then, for all $(s,t) \in \triangle_T$, $X_{s,t} = \left(1, X_{s,t}^1\right)$ with $X_{s,t}^1 \in V$. For all $0 \leq s \leq u \leq t \leq T$, we have $X_{s,t} = X_{s,u} \otimes X_{u,t}$, so that

$$(1, X_{s,t}^1) = (1, X_{s,u}^1) \otimes (1, X_{u,t}^1) = (1, X_{s,u}^1 + X_{u,t}^1).$$

Thus, the multiplicative condition in $T^{(1)}(V)$ reduces to the additivity in V of the mapping $(s,t) \mapsto X_{s,t}^1$, that is, to the identity $X_{s,t}^1 = X_{s,u}^1 + X_{u,t}^1$.

This identity is equivalent to the existence of a path $x : [0,T] \longrightarrow V$ such that $X_{s,t}^1 = x_t - x_s$ for all $(s,t) \in \triangle_T$. The path x is uniquely determined by X up to the addition of a constant element of V. Finally, a multiplicative functional of degree 1 in V is just a path in V modulo translation described through its increments.

Example 3.3. Let us now consider a multiplicative functional X of degree 2, so that for all $(s,t) \in \triangle_T$, $X_{s,t} \in T^{(2)}(V)$. According to the rules defining the tensor product in $T^{(2)}(V)$ (see Definition 2.5), the identities satisfied by $X_{s,t}$ are the following: for all $0 \leq s \leq u \leq t \leq T$,

$$(\text{level } 1) \ \ X_{s,t}^1 = X_{s,u}^1 + X_{u,t}^1,$$

$$(\text{level } 2) \ \ X_{s,t}^2 = X_{s,u}^2 + X_{u,t}^2 + X_{s,u}^1 \otimes X_{u,t}^1.$$

We will come back to these relations later in this chapter and in particular give a geometrical interpretation of the second one in terms of area.

In general, if X is a multiplicative functional of degree n in V, then $\pi_1 \circ X : \triangle_T \longrightarrow T^{(1)}(V)$ is a multiplicative functional of degree 1. Hence, by the remark made earlier, there exists a classical path $x : [0, T] \longrightarrow V$ which underlies X in the sense that $X_{s,t}^1 = x_t - x_s$ for all $(s, t) \in \triangle_T$. However, X is not at all the signature of x in general. For one thing, x may not have finite p-variation for any $p < 2$, in which case the signature of x does not exist, and even if x has finite p-variation for some $p < 2$, X may be different from the signature of X. For instance, if w is any non-zero tensor in $V^{\otimes 2}$, then the functional $X_{s,t} = (1, 0, (t - s)w) \in T^{(2)}(V)$ is a multiplicative functional of degree 2. However, the path x determined by $\pi_1 \circ X$ is a constant path and the signature of a constant path truncated at level 2 is simply $(1, 0, 0)$. More generally, if $\psi : [0, T] \longrightarrow V^{\otimes n}$ is any function, then it is not hard to check that the functional $X_{s,t} = (1, 0, \ldots, 0, \psi(t) - \psi(s)) \in T^{(n)}(V)$ is multiplicative. This is a particular instance of the following simple and very useful algebraic lemma.

Lemma 3.4. *Let $m \geq 0$ be an integer. Let $X, Y : \triangle_T \longrightarrow T^{(m+1)}(V)$ be two multiplicative functionals which agree up to the m-th degree, that is, such that $\pi_m(X) = \pi_m(Y)$ or in other words such that $X_{s,t}^i = Y_{s,t}^i$ for $i = 0, \ldots, m$. Then the difference function $\Psi : \triangle_T \longrightarrow V^{\otimes m+1}$ defined by:*

$$\Psi_{s,t} = X_{s,t}^{m+1} - Y_{s,t}^{m+1} \in V^{\otimes m+1}$$

is additive, which means that for all $s \leq u \leq t$,

$$\Psi_{s,t} = \Psi_{s,u} + \Psi_{u,t}.$$

Conversely, if $X : \triangle_T \longrightarrow T^{(m+1)}(V)$ is a multiplicative functional and if $\Psi : \triangle_T \longrightarrow V^{\otimes m+1}$ is an additive function, then $(s, t) \mapsto X_{s,t} + \Psi_{s,t}$ is still a multiplicative functional.

In the last statement, we have implicitly used the imbedding of $V^{\otimes m+1}$ into $T^{(m+1)}(V)$ to add $X_{s,t}$ and $\Psi_{s,t}$. The proof of this lemma is easy and it is left as an exercise for the reader.

Remark 3.5. Let X be a multiplicative functional of degree n in V. It is possible to see X as a path in $T^{(n)}(V)$ by considering $t \mapsto X_{0,t}$. It turns out that the path $t \mapsto X_{0,t}$ characterises X completely. Indeed, for every $t \in [0, T]$, the term of degree 0 of $X_{0,t}$ is equal to 1, so that $X_{0,t}$ is invertible in $T^{(n)}(V)$. Hence, Chen's identity gives, for all $(s, t) \in \triangle_T$, $X_{s,t} = (X_{0,s})^{-1} \otimes X_{0,t}$.

Stochastic integration gives rise to iterated integrals which are not in the scope of the Young integral. It is not surprising that it provides us with very important examples of multiplicative functionals. The Itô and Stratonovich signatures of a Brownian path are two of them. Let $(B_t)_{t \geq 0}$ be a standard

Brownian motion in $\mathbb{R}^d$. Then its Itô and Stratonovich signatures are respectively defined, outside a negligible set, by

$$S^{It\hat{o}}(B)_{s,t} = \left(1, B_t - B_s, \iint_{s<u_1<u_2<t} dB_{u_1} \otimes dB_{u_2}, \ldots \right)$$

$$S^{Strat}(B)_{s,t} = \left(1, B_t - B_s, \iint_{s<u_1<u_2<t} \circ dB_{u_1} \otimes \circ dB_{u_2}, \ldots \right).$$

For example, if $e_1, \ldots, e_d$ is the canonical basis of $\mathbb{R}^d$ and $B^1, \ldots, B^d$ are the corresponding components of the Brownian motion, then the term of degree 2 in the Itô signature of B is

$$S^{It\hat{o}}(B)^2_{s,t} = \sum_{i,j=1}^{d} \int_{s<u_1<u_2<t} dB^i_{u_1} dB^j_{u_2} \; e_i \otimes e_j.$$

The fact that these functionals are multiplicative is a consequence of the properties of the Itô and Stratonovich integrals.

So far, the only condition imposed on a functional X has been Chen's identity (3.1), which is a purely algebraic condition. We are now going to introduce a notion of p-variation for multiplicative functionals. When x is a positive real number, we use the notation $x! = \Gamma(x+1)$. Recall the definition of a control (Definition 1.9).

Definition 3.6. *Let $p \geq 1$ be a real number and $n \geq 1$ be an integer. Let $\omega : [0,T] \longrightarrow [0,+\infty)$ be a control. Let $X : \triangle_T \longrightarrow T^{(n)}(V)$ be a multiplicative functional.*

We say that X has finite p-variation on $\triangle_T$ controlled by ω if

$$\left\| X^i_{s,t} \right\| \leq \frac{\omega(s,t)^{\frac{i}{p}}}{\beta \left(\frac{i}{p} \right)!} \quad \forall i = 1 \ldots n , \;\; \forall (s,t) \in \triangle_T.$$

In general, we say that X has finite p-variation if there exists a control ω such that the conditions above are satisfied.

In this definition, β is a real constant which depends only on p and will be explicited later. The precise value of β does not affect the class of multiplicative functionals with finite p-variation, nor does the factor $\left(\frac{i}{p} \right)!$, since we only put conditions on X^i for finitely many values of i. Nevertheless, the presence of this factor and an appropriate choice of β will allow us to simplify the numerical constants appearing in many of the forthcoming theorems.

3.1.2 Extension of Multiplicative Functionals

Lemma 3.4 shows that a multiplicative functional is never completely determined by its lower degree terms. More precisely, given any multiplicative functional of degree $n \geq 1$, there exists another functional, distinct from the first one but such that both agree up to the level $n-1$. However, if both functionals have finite p-variation, then the definition of the p-variation of a multiplicative functional shows that the difference at level n between the two functionals, denoted in Lemma 3.4 by Ψ, is of the form $\Psi_{s,t} = \psi(s) - \psi(t)$ for some ψ which is Hölder continuous of exponent $\frac{n}{p}$. If $n > p$, then ψ is necessarily constant. This short argument shows that *within the class of functionals with finite p-variation, a multiplicative functional is determined by its truncature at level $\lfloor p \rfloor$*.

For instance, let $x : [0,T] \longrightarrow V$ be a path with finite p-variation for some $p < 2$. Then the signature of x, denoted by X, is the unique multiplicative functional *with finite p-variation* such that $\pi_1(X_{s,t}) = (1, x_t - x_s)$ for all $(s,t) \in \triangle_T$.

In the other direction, let X be a multiplicative functional of degree $n \geq 1$ with finite p-variation for some $p < 2$. Let x be the path in V underlying $\pi_1 \circ X$. Then the argument earlier shows that X coincides with the signature of x truncated at level n. It is now very easy to extend X to a multiplicative functional of arbitrarily high degree, by adding higher order iterated integrals of x. Moreover, we know that this extension is continuous with respect to X, because the iterated integrals depend continuously on x.

The theorem which we are now going to prove, which is the first fundamental result in the theory of rough paths, generalises the remarks made earlier and states that every multiplicative functional of degree n and with finite p-variation can be extended in a unique way to a multiplicative functional of arbitrary high degree, provided n is greater than the integer part of p. Furthermore, the extension map thus defined is continuous in the p-variation metric.

Theorem 3.7 (Extension Theorem). *Let $p \geq 1$ be a real number and $n \geq 1$ an integer. Let $X : \triangle_T \longrightarrow T^{(n)}(V)$ be a multiplicative functional with finite p-variation controlled by a control ω. Assume that $n \geq \lfloor p \rfloor$. Then there exists a unique extension of X to a multiplicative functional $\triangle_T \longrightarrow T((V))$ which possesses finite p-variation.*

More precisely, for every $m \geq \lfloor p \rfloor + 1$, there exists a unique continuous function $X^m : \triangle_T \longrightarrow V^{\otimes m}$ such that

$$(s,t) \mapsto X_{s,t} = (1, X_{s,t}^1, \dots, X_{s,t}^{\lfloor p \rfloor}, \dots, X_{s,t}^m, \dots) \in T((V))$$

is a multiplicative functional with finite p-variation controlled by ω. By this we mean that

$$\left\| X_{s,t}^i \right\| \leq \frac{\omega(s,t)^{\frac{i}{p}}}{\beta \left(\frac{i}{p} \right)!} \quad \forall i \geq 1, \ \forall (s,t) \in \triangle_T,$$

where

$$\beta = p^2 \left(1 + \sum_{r=3}^{\infty} \left(\frac{2}{r-2}\right)^{\frac{\lfloor p \rfloor + 1}{p}}\right).$$

Just as the proof of the existence of the Young integral (Theorem 1.16), our proof of the extension theorem relies on a maximal inequality which we establish by progressively coarsening a partition in a clever way. Before that, we need to state a generalisation of the binomial theorem which we call the neo-classical inequality.

Lemma 3.8 (Neo-classical Inequality). *For any* $p \in [1, \infty)$, $n \in \mathbb{N}$ *and* $s, t \geq 0$,

$$\frac{1}{p^2} \sum_{i=0}^{n} \frac{s^{\frac{i}{p}} t^{\frac{n-i}{p}}}{\left(\frac{i}{p}\right)! \left(\frac{n-i}{p}\right)!} \leq \frac{(s+t)^{\frac{n}{p}}}{\left(\frac{n}{p}\right)!}.$$

This inequality is of course an equality when $p = 1$. It is strongly conjectured, and supported by numerical simulations, that the inequality is still true when p^2 is replaced by p in the left-hand side.

The best known proof of this lemma is not nearly as simple as one might expect at first sight. It relies on the maximum principle for parabolic PDE's. We do not offer it in these notes, but refer the reader to [60] and [67].

Using this inequality, we can understand the particular choice of coefficients in definition (3.6) of p-variation. Suppose that $a = (1, a_1, \ldots a_{n-1}, 0) \in T^{(n)}(V)$ and $b = (1, b_1, \ldots b_{n-1}, 0) \in T^{(n)}(V)$. Assume that for each $i = 1, 2, \ldots, n-1$,

$$\|a_i\| \leq \frac{s^{\frac{i}{p}}}{\beta \left(\frac{i}{p}\right)!} \qquad \text{and} \qquad \|b_i\| \leq \frac{t^{\frac{i}{p}}}{\beta \left(\frac{i}{p}\right)!}.$$

Then, if we set $c = a \otimes b$ (multiplication in $T^{(n)}(V)$), we have

$$\|c_n\| \leq \left\|\sum_{i=0}^{n} a_i \otimes b_{n-i}\right\| \leq \sum_{i=0}^{n} \|a_i\| \|b_{n-i}\|$$

$$\leq \sum_{i=0}^{n} \frac{s^{\frac{i}{p}} t^{\frac{n-i}{p}}}{\beta \left(\frac{i}{p}\right)! \beta \left(\frac{n-i}{p}\right)!}$$

$$\leq \frac{p^2}{\beta^2} \frac{(s+t)^{\frac{n}{p}}}{\left(\frac{n}{p}\right)!}.$$

Exercise 3.9. Check that the neo-classical inequality has the following multinomial extension: for any $p \in [1, \infty)$, $r, n \in \mathbb{N}^*$ and $x_1, x_2, \ldots x_r \geq 0$,

$$\frac{1}{p^{2r-2}} \sum_{\substack{k_1,\dots k_r \in \mathbb{N} \\ k_1+\dots+k_r=n}} \frac{x_1^{\frac{k_1}{p}}}{\left(\frac{k_1}{p}\right)!} \cdots \frac{x_r^{\frac{k_r}{p}}}{\left(\frac{k_r}{p}\right)!} \leq \frac{(x_1+\dots+x_r)^{\frac{n}{p}}}{\left(\frac{n}{p}\right)!}.$$

We are now ready to prove Theorem 3.7.

Proof (Extension Theorem).

Uniqueness – Suppose that $X^{(n+1)}$ and $\widetilde{X}^{(n+1)}$ are two extensions of $X^{(n)}$ to $T^{(n+1)}(V)$. We know that both $X^{(n+1)}$ and $\widetilde{X}^{(n+1)}$ have finite p-variation controlled by the same ω. Thus, using the fact that $X^{(n+1)}$ and $\widetilde{X}^{(n+1)}$ agree up to the nth level, we find

$$\|X_{s,t}^{(n+1)} - \widetilde{X}_{s,t}^{(n+1)}\| = \|X_{s,t}^{n+1} - \widetilde{X}_{s,t}^{n+1}\| \leq C\omega(s,t)^{\frac{n+1}{p}}$$

for some constant C. Let us denote by Ψ the difference between $X^{(n+1)}$ and $\widetilde{X}^{(n+1)}$. It follows from Lemma 3.4 that the function $(s,t) \mapsto \Psi_{s,t}$ is additive in $V^{\otimes(n+1)}$, so that $t \mapsto \Psi_{0,t}$ is a path of finite $\frac{n+1}{p}$-variation in $V^{\otimes(n+1)}$ starting at zero. We conclude from the regularity of ω and the fact that $\frac{n+1}{p} > 1$ that Ψ is identically equal to zero. The uniqueness of the extension follows.

Existence – We prove the existence of the extension X^m by induction on m. More precisely, for each $m \in \{n, n+1, \dots\}$ we prove the following statement.

Extending to level $m+1$ – Let $X^{(m)} : \triangle_T \longrightarrow T^{(m)}(V)$ be a multiplicative functional whose p-variation is controlled by ω, that is, such that $X^{(m)}$ satisfies, for all $0 \leq s \leq t \leq u \leq T$:

1. $X_{s,t}^{(m)} = \left(1, X_{s,t}^1, \dots X_{s,t}^m\right) \in T^{(m)}(V)$
2. $X_{s,u}^{(m)} = X_{s,t}^{(m)} \otimes X_{t,u}^{(m)}$
3. $\left\|X_{s,t}^i\right\| \leq \frac{\omega(s,t)^{\frac{i}{p}}}{\beta\left(\frac{i}{p}\right)!}$ for all $i \leq m$

Then there exists a multiplicative functional $X^{(m+1)} : \triangle_T \longrightarrow T^{(m+1)}(V)$ *which also satisfies these three conditions, and agrees with* $X^{(m)}$ *up to level m.*

The first thing that we need to do is to construct a functional with values in $T^{(m+1)}(V)$. We do this in the following way. For all $(s,t) \in \triangle_T$, set

$$\widehat{X}_{s,t} = \left(1, X_{s,t}^1, \dots X_{s,t}^m, 0\right) \in T^{(m+1)}(V). \tag{3.2}$$

In general, this functional is not multiplicative, but at least it takes values in $T^{(m+1)}(V)$. We claim that the multiplicative functional we want to construct is given by:

$$X_{s,t}^{(m+1)} = \lim_{|\mathcal{D}| \to 0} \widehat{X}_{s,t}^{\mathcal{D}}$$

where for each partition $\mathcal{D} = \{s = t_0 \leq t_1 \leq \dots \leq t_r = t\}$ of $[s,t]$ we define

$$\widehat{X}_{s,t}^{\mathcal{D}} = \widehat{X}_{s,t_1} \otimes \dots \otimes \widehat{X}_{t_{r-1},t}$$

using the multiplication in $T^{(m+1)}(V)$.

Before we prove that this limit actually exists, observe that if it does exist, then it will surely be multiplicative. Indeed, let $u \in (s,t)$ and choose a sequence of partitions $\mathcal{D}_n$ of $[s,t]$ such that $|\mathcal{D}_n| \to 0$ as $n \to \infty$. Set $\widetilde{\mathcal{D}}_n = \mathcal{D}_n \cup \{u\}$. Then we still have $|\widetilde{\mathcal{D}}_n| \to 0$ as $n \to \infty$ and

$$
\begin{aligned}
X_{s,t}^{(m+1)} &= \lim_{|\mathcal{D}| \to 0} \widehat{X}_{s,t}^{\mathcal{D}} = \lim_{n \to \infty} \widehat{X}_{s,t}^{\widetilde{\mathcal{D}}_n} \\
&= \lim_{n \to \infty} \widehat{X}_{s,u}^{\widetilde{\mathcal{D}}_n \cap [s,u]} \otimes \widehat{X}_{u,t}^{\widetilde{\mathcal{D}}_n \cap [u,t]} \\
&= \left(\lim_{n \to \infty} \widehat{X}_{s,u}^{\widetilde{\mathcal{D}}_n \cap [s,u]} \right) \otimes \left(\lim_{n \to \infty} \widehat{X}_{u,t}^{\widetilde{\mathcal{D}}_n \cap [u,t]} \right) \\
&= X_{s,u}^{(m+1)} \otimes X_{u,t}^{(m+1)}.
\end{aligned}
$$

In order to complete the induction argument, we thus have to prove that this limit exists and that its p-variation is controlled by ω. Observe that in both cases the difficulty lies in understanding the terms of degree $m + 1$ in $\widehat{X}_{s,t}^{\mathcal{D}}$, because $X_{s,t}^{(m)}$ is multiplicative and so, for $i = 0, \dots, m$, $(\widehat{X}_{s,t}^{\mathcal{D}})^i = X_{s,t}^i$. Estimating the magnitude of this limit actually comes before proving its existence.

The heart of our argument is the following maximal inequality. We claim that under the induction hypothesis, for any partition $\mathcal{D}$ of $[s,t]$, we have:

$$
\left\| (\widehat{X}_{s,t}^{\mathcal{D}})^i \right\| \leq \frac{\omega\,(s,t)^{\frac{i}{p}}}{\beta \left(\frac{i}{p} \right)!} \quad \text{for all } (s,t) \in \Delta_T \text{ and } i \leq m + 1. \tag{3.3}
$$

By the remark made a few lines earlier, this inequality follows at once from the induction hypothesis for $i = 0, \dots, m$. Now let us fix a partition $\mathcal{D} = \{s = t_0 < \dots < t_r = t\}$. For any other partition $\mathcal{D}'$, we have:

$$
\left\| (\widehat{X}_{s,t}^{\mathcal{D}})^{m+1} \right\| \leq \left\| (\widehat{X}_{s,t}^{\mathcal{D}} - \widehat{X}_{s,t}^{\mathcal{D}'})^{m+1} \right\| + \left\| (\widehat{X}_{s,t}^{\mathcal{D}'})^{m+1} \right\|.
$$

Just as in the construction of the Young integral (Theorem 1.16), we are going to progressively coarsen our fixed partition $\mathcal{D}$ by dropping carefully chosen points. Let $\mathcal{D}' = \mathcal{D} \setminus \{t_j\}$ where t_j is t_1 if $r = 2$ and is otherwise chosen such that

$$
\omega(t_{j-1}, t_{j+1}) \leq \frac{2}{r-1} \omega(s,t).
$$

This is possible, exactly as in the proof of Theorem 1.16. Then

$$
\begin{aligned}
\widehat{X}_{s,t}^{\mathcal{D}} - \widehat{X}_{s,t}^{\mathcal{D}'} &= \widehat{X}_{s,t_1} \otimes \dots \otimes \widehat{X}_{t_{j-2},t_{j-1}} \otimes \left(\widehat{X}_{t_{j-1},t_j} \otimes \widehat{X}_{t_j,t_{j+1}} - \widehat{X}_{t_{j-1},t_{j+1}} \right) \\
&\otimes \widehat{X}_{t_{j+1},t_{j+2}} \otimes \dots \otimes \widehat{X}_{t_{r-1},t}.
\end{aligned} \tag{3.4}
$$

Now, it follows from the definition (3.2) of $\widehat{X}$ and the multiplicative nature of $X^{(m)}$ that $\widehat{X}_{t_{j-1},t_j} \otimes \widehat{X}_{t_j,t_{j+1}} - \widehat{X}_{t_{j-1},t_{j+1}}$ is a pure $(m+1)$-tensor:

$$\widehat{X}_{t_{j-1},t_j} \otimes \widehat{X}_{t_j,t_{j+1}} - \widehat{X}_{t_{j-1},t_{j+1}} = \left(0,\ldots,0,\sum_{i=1}^{m} X_{t_{j-1},t_j}^{i} \otimes X_{t_j,t_{j+1}}^{m+1-i}\right).$$

Thanks to this property and to the fact that we are computing in $T^{(m+1)}(V)$, the only non-zero terms in (3.4) is obtained by picking the constant terms of $\widehat{X}_{s,t_1},\ldots,\widehat{X}_{t_{j-2},t_{j-1}}$, then the term of degree $m+1$ in the central term, and finally again the constant terms in $\widehat{X}_{t_{j+1},t_{j+2}},\ldots,\widehat{X}_{t_{r-1},t}$. Hence, the difference $\widehat{X}_{s,t}^{\mathcal{D}} - \widehat{X}_{s,t}^{\mathcal{D}'}$ has the following simple expression:

$$\widehat{X}_{s,t}^{\mathcal{D}} - \widehat{X}_{s,t}^{\mathcal{D}'} = \left(0,\ldots,0,\sum_{i=1}^{m} X_{t_{j-1},t_j}^{i} \otimes X_{t_j,t_{j+1}}^{m+1-i}\right).$$

In particular,

$$\left\|(\widehat{X}_{s,t}^{\mathcal{D}} - \widehat{X}_{s,t}^{\mathcal{D}'})^{m+1}\right\| = \left\|\sum_{i=1}^{m} X_{t_{j-1},t_j}^{i} \otimes X_{t_j,t_{j+1}}^{m+1-i}\right\| \le \sum_{i=1}^{m}\left\| X_{t_{j-1},t_j}^{i}\right\|\left\| X_{t_j,t_{j+1}}^{m+1-i}\right\|.$$

From the induction hypothesis, we get:

$$\sum_{i=1}^{m}\left\| X_{t_{j-1},t_j}^{i}\right\|\left\| X_{t_j,t_{j+1}}^{m+1-i}\right\| \le \sum_{i=1}^{m}\left(\frac{\omega(t_{j-1},t_j)^{\frac{i}{p}}}{\beta\left(\frac{i}{p}\right)!}\right)\left(\frac{\omega(t_j,t_{j+1})^{\frac{m+1-i}{p}}}{\beta\left(\frac{m+1-i}{p}\right)!}\right)$$

and so, by the neo-classical inequality (Lemma 3.8), the super-additivity of ω and the choice of t_j,

$$\sum_{i=1}^{m}\left\| X_{t_{j-1},t_j}^{i}\right\|\left\| X_{t_j,t_{j+1}}^{m+1-i}\right\| \le \frac{p^2}{\beta^2}\frac{(\omega(t_{j-1},t_j)+\omega(t_j,t_{j+1}))^{\frac{m+1}{p}}}{\left(\frac{m+1}{p}\right)!}$$

$$\le \frac{p^2}{\beta^2}\frac{\omega(t_{j-1},t_{j+1})^{\frac{m+1}{p}}}{\left(\frac{m+1}{p}\right)!}$$

$$\le \frac{p^2}{\beta}\left(\frac{2}{r-1}\right)^{\frac{m+1}{p}}\frac{\omega(s,t)^{\frac{m+1}{p}}}{\beta\left(\frac{m+1}{p}\right)!}.$$

By successively dropping points from $\mathcal{D}$ until $\mathcal{D} = \{s,t\}$, we get

$$\left\|(\widehat{X}_{s,t}^{\mathcal{D}} - \widehat{X}_{s,t})^{m+1}\right\| \le \frac{p^2}{\beta}\left(1+2^{\frac{m+1}{p}}\left(\zeta\left(\frac{\lfloor p\rfloor+1}{p}\right)-1\right)\right)\frac{\omega(s,t)^{\frac{m+1}{p}}}{\beta\left(\frac{m+1}{p}\right)!}$$

where $\zeta(z)=\sum_{n\ge 1}^{\infty} n^{-z}$. Therefore if we choose

$$\beta \ge p^2\left(1+2^{\frac{m+1}{p}}\left(\zeta\left(\frac{\lfloor p\rfloor+1}{p}\right)-1\right)\right),$$

then we have the required estimate, namely

$$\left\|(\widehat{X}_{s,t}^{\mathcal{D}})^{m+1}\right\| \leq \frac{\omega\left(s,t\right)^{\frac{m+1}{p}}}{\beta\left(\frac{m+1}{p}\right)!}.$$

This estimate holds for every partition $\mathcal{D}$ of $[s,t]$ and this completes the proof of the maximal inequality (3.3).

We still have to prove the existence of the limit $\lim_{|\mathcal{D}|\to 0} \widehat{X}_{s,t}^{\mathcal{D}}$. We are going to show that the family $\widehat{X}_{s,t}^{\mathcal{D}}$ satisfies a Cauchy convergence principle. Suppose that $\mathcal{D}$ and $\widetilde{\mathcal{D}}$ are two partitions of $[s,t]$, both having mesh size less than δ. Let $\widehat{\mathcal{D}}$ be a common refinement of these two partitions. Fix an interval $[t_j, t_{j+1}]$ in $\mathcal{D}$. Then the refinement $\widehat{\mathcal{D}}$ breaks up $[t_j, t_{j+1}]$ in a number of pieces, and we get a partition of $[t_j, t_{j+1}]$, which we denote by $\widehat{\mathcal{D}}_j = \widehat{\mathcal{D}} \cap [t_j, t_{j+1}]$.

The maximal inequality allows us to estimate $(\widehat{X}_{t_j,t_{j+1}}^{\widehat{\mathcal{D}}_j} - \widehat{X}_{t_j,t_{j+1}})^{m+1}$ and all terms of degree less than $m+1$ in this difference are zero. Therefore,

$$\left\|\widehat{X}_{t_j,t_{j+1}}^{\widehat{\mathcal{D}}_j} - \widehat{X}_{t_j,t_{j+1}}\right\| \leq \frac{\omega\left(t_j,t_{j+1}\right)^{\frac{m+1}{p}}}{\beta\left(\frac{m+1}{p}\right)!}.$$

On the other hand,

$$\widehat{X}_{s,t}^{\widehat{\mathcal{D}}} - \widehat{X}_{s,t} = \widehat{X}^{\widehat{\mathcal{D}}_0} \otimes \ldots \otimes \widehat{X}^{\widehat{\mathcal{D}}_{r-1}} - \widehat{X}_{t_0,t_1} \otimes \ldots \otimes \widehat{X}_{t_{r-1},t_r}$$

$$= \sum_{j=0}^{r-1} \widehat{X}^{\widehat{\mathcal{D}}_0} \otimes \ldots \otimes \widehat{X}^{\widehat{\mathcal{D}}_{j-1}} \otimes (\widehat{X}^{\widehat{\mathcal{D}}_j} - \widehat{X}_{t_{j-1},t_j})$$

$$\otimes \widehat{X}_{t_j,t_{j+1}} \otimes \ldots \otimes \widehat{X}_{t_{r-1},t_r}.$$

We have seen earlier that, for each j, the term between the brackets is a pure $(m+1)$-tensor. All levels but the $(m+1)$-th vanish in the sum and the $(m+1)$-th is equal to the sum of those appearing between the brackets. Thus,

$$\left\|\widehat{X}_{s,t}^{\widehat{\mathcal{D}}} - \widehat{X}_{s,t}^{\mathcal{D}}\right\| \leq \sum_{\mathcal{D}} \frac{\omega\left(t_j,t_{j+1}\right)^{\frac{m+1}{p}}}{\beta\left(\frac{m+1}{p}\right)!}$$

$$\leq \frac{\omega\left(s,t\right)}{\beta\left(\frac{m+1}{p}\right)!} \cdot \left[\max_{|u-v|<\delta} \omega\left(u,v\right)\right]^{\frac{m+1}{p}-1}.$$

This is independent of the refinement $\widehat{\mathcal{D}}$ and since $m+1 > p$ and ω is uniformly continuous, the right-hand side can be made arbitrarily small, uniformly over all partitions with mesh size smaller than δ. Applying the triangle inequality, we have a uniform bound on $X^{\mathcal{D}} - X^{\widetilde{\mathcal{D}}}$.

We can therefore conclude that $X_{s,t}^{(m+1)} = \lim_{|\mathcal{D}|\to 0} \widehat{X}_{s,t}^{\mathcal{D}}$ exists and satisfies all the requirements of the induction. $\square$

The extension theorem tells us that for a given multiplicative functional X of finite p-variation, the terms X^k for $k > \lfloor p \rfloor$, are uniquely determined by the lower order terms X^i for $i \leq \lfloor p \rfloor$. On the other hand, it is not difficult to see that the terms X^i for $i \leq \lfloor p \rfloor$ are never uniquely determined by the terms of lower order. As an example, consider the multiplicative functional defined by the Stratonovich signature of d-dimensional Brownian motion, up to level 2, i.e.

$$S^{Strat}(B)_{s,t} = \left(1, B_t - B_s, \iint_{s<u_1<u_2<t} \circ dB_{u_1} \otimes \circ dB_{u_2}\right)$$

We can define a new functional X by,

$$X_{s,t} = \left(1, B_t - B_s, \iint_{s<u_1<u_2<t} \circ dB_{u_1} \otimes \circ dB_{u_2} + \frac{t-s}{2}I_d\right),$$

where I_d is the tensor $\sum_{i=1}^d e_i \otimes e_i$ with $(e_1,\ldots,e_d)$ the canonical basis of $\mathbb{R}^d$. Then X is in fact equal to the Itô signature of Brownian motion up to level 2, and hence we know that it is multiplicative. This is just a special case of the following more general situation. Suppose we are given a multiplicative functional $X_{s,t} = \left(1, X_{s,t}^1, X_{s,t}^2\right)$ with finite 2-variation. Then for any path $\psi : [0,T] \longrightarrow V^{\otimes 2}$ with bounded variation, $\widetilde{X}_{s,t} = \left(1, X_{s,t}^1,, X_{s,t}^2 + \psi(t) - \psi(s)\right)$ is again a multiplicative functional of finite 2-variation.

3.1.3 Continuity of the Extension Map

We have just shown that if we impose an appropriate p-variation condition on a multiplicative functional, then the higher order terms of this functional are completely determined by the lower order terms. A natural question to ask now is whether this extension map is continuous in the p-variation metric. Given two multiplicative functionals of finite p-variation in $T^{(n)}(V)$, $n \geq \lfloor p \rfloor$, which are close in the p-variation metric, are the respective extensions also close?

This question is answered positively in the following theorem, which furthermore gives a very explicit estimate for the modulus of continuity of the extension map.

Theorem 3.10. *Let X and Y be two multiplicative functionals in $T^{(n)}(V)$ of finite p-variation, $n \geq \lfloor p \rfloor$, controlled by ω. Suppose that for some $\varepsilon \in (0,1)$,*

$$\left\|X_{s,t}^i - Y_{s,t}^i\right\| \leq \varepsilon \frac{\omega(s,t)^{\frac{i}{p}}}{\beta\left(\frac{i}{p}\right)!} \tag{3.5}$$

for $i = 1, \ldots, n$ and for all $(s,t) \in \triangle_T$. Then, provided β is chosen such that

$$\beta \geq 2p^2 \left(1 + \sum_{r=3}^{\infty} \left(\frac{2}{r-2} \right)^{\frac{\lfloor p \rfloor + 1}{p}} \right),$$

the control (3.5) holds for all i.

The proof of this theorem is very similar to that of the extension theorem. The details of the proof can be found in [60] and [67].

3.2 Spaces of Rough Paths

3.2.1 Rough Paths and the p-Variation Topology

We can now identify the elementary classes of objects that will drive differential equations in the following chapters.

Definition 3.11 (Rough Path). *Let V be a Banach space. Let $p \geq 1$ be a real number. A p-rough path in V is a multiplicative functional of degree $\lfloor p \rfloor$ in V with finite p-variation. The space of p-rough paths is denoted by $\Omega_p(V)$.*

Thus, a p-rough path is a continuous mapping from $\triangle_T$ to $T^{(\lfloor p \rfloor)}(V)$ which satisfies an algebraic condition, Chen's identity, and an analytic condition, the finiteness of its p-variation. It follows from the extension theorem (Theorem 3.7) that every p-rough path $X \in \Omega_p(V)$ has a full signature, that is, it can be extended to a multiplicative functional of arbitrarily high degree with finite p-variation in V.

When $p \geq 2$, the space $\Omega_p(V)$ is *not* a vector space, because of the non-linearity of Chen's identity (3.1). There is nothing like the sum or the difference of two rough paths. We will come back to this point in great detail in Chap. 5. The space $\Omega_p(V)$ is nevertheless a cone, that is, one can multiply a rough path by a real constant. One can also add to a rough path a path with bounded variation. See [67, Sect. 3.3.3] for more details.

Let $C_{0,p}\left(\triangle_T, T^{(\lfloor p \rfloor)}(V)\right)$ be the space of all continuous functions from the simplex $\triangle_T$ into the truncated tensor algebra $T^{(\lfloor p \rfloor)}(V)$ with finite p-variation. We define the p-variation metric on this linear space as follows: for $X, Y \in C_{0,p}\left(\triangle_T, T^{(\lfloor p \rfloor)}(V)\right)$, set

$$d_p(X,Y) = \max_{1 \leq i \leq \lfloor p \rfloor} \sup_{\mathcal{D} \subset [0,T]} \left(\sum_{\mathcal{D}} \| X^i_{t_{l-1},t_l} - Y^i_{t_{l-1},t_l} \|^{\frac{p}{i}} \right)^{\frac{1}{p}}.$$

The space $\Omega_p(V)$ is a subspace of $C_{0,p}\left(\triangle_T, T^{\lfloor p \rfloor}(V)\right)$ and if we equip it with the p-variation metric d_p, then $(\Omega_p(V), d_p)$ becomes a complete metric space (see [67, Sect. 3.3.1]). In practice, it is often quite difficult to work with this metric. Fortunately, it is possible to find a nice characterisation of the notion of convergence induced by d_p.

Definition 3.12. *A sequence* $\{X(n)\}_{n\in\mathbb{N}}$ *of elements of* $C_{0,p}\left(\triangle_T, T^{(\lfloor p \rfloor)}(V)\right)$ *is said to converge to* $X \in C_{0,p}\left(\triangle_T, T^{\lfloor p \rfloor}(V)\right)$ *in the* p-*variation topology if there exists a control* ω *and a sequence* $(a(n))_{n\geq 0}$ *such that* $\lim_{n\to\infty} a(n) = 0$ *and, for all* $i = 0, \ldots, p$, *all* $(s,t) \in \triangle_T$ *and all* $n \geq 0$,

$$\left\| X(n)_{s,t}^i \right\|, \left\| X_{s,t}^i \right\| \leq \omega(s,t)^{\frac{i}{p}} \quad and \quad \left\| X(n)_{s,t}^i - X_{s,t}^i \right\| \leq a(n)\,\omega(s,t)^{\frac{i}{p}}.$$

It can be shown that if $d_p(X(n), X) \to 0$, then $X(n)$ converges to X in the sense of this definition. The converse is almost true: from a sequence which converges in the p-variation topology, one can always extract a sub-sequence which converges for the distance d_p (see [67]).

3.2.2 Geometric Rough Paths

Recall that a path in V with finite q-variation for some $q < 2$ defines a q-rough path. Indeed, a path determines a multiplicative functional of degree $1 = \lfloor q \rfloor$ with finite q-variation. Given $p \geq q$, this q-rough path can be extended to a multiplicative functional of degree $\lfloor p \rfloor$ with finite q-variation, hence finite p-variation. In particular, a path with bounded variation can be considered canonically as a p-rough path for every $p \geq 1$.

Definition 3.13. *A geometric* p-*rough path is a* p-*rough path that can be expressed as a limit of* 1-*rough paths in the* p-*variation distance.*
   The space of geometric p-*rough paths in* V *is denoted by* $G\Omega_p(V)$.

Thus the space of geometric p-rough paths in V is the closure in $(\Omega_p(V), d_p)$ of the space $\Omega_1(V) = \mathcal{V}^1([0,T], V)$ of paths with bounded variation.

Recall that the signature of a path with finite p-variation for some $p < 2$, truncated at level n, takes its values in the free nilpotent group of step n, denoted by $G^{(n)}$ (see Sect. 2.2.5). Since $G^{(n)}$ is defined by algebraic identities, namely identities related to the shuffle product, it is still true that a geometric p-rough path takes its values in $G^{(\lfloor p \rfloor)}$. Unfortunately, this does not characterise geometric rough paths among all rough paths. This makes the following definition necessary and meaningful.

Definition 3.14. *A weakly geometric* p-*rough path is a* p-*rough path which takes its values in* $G^{(\lfloor p \rfloor)}$, *the free nilpotent group of step* $\lfloor p \rfloor$.
   The space of weakly geometric rough paths is denoted by $WG\Omega_p(V)$.

The following inclusions hold:

$$G\Omega_p(V) \subset WG\Omega_p(V) \subset \Omega_p(V).$$

All inclusions are strict, although this is not at all obvious for the first one (see [12]). The difference between geometric and weakly geometric rough paths is indeed annoying and generally insignificant. It could be compared to the difference between C^1 and Lipschitz functions. On the other hand, the difference between weakly geometric and plain rough paths is substantial. It could be compared with the difference between flows of diffeomorphisms and time-dependent differential operators, or between the elements of a Lie group and those of its enveloping Lie algebra.

Exercise 3.15. Let X be a weakly geometric rough path in V. Check that, for every $n \geq 1$, the symmetric part of $X_{s,t}^n$ is equal to $(X_{s,t}^1)^{\otimes n}/n!$. For this, use the definition of group-like elements and the fact that, if e^* belongs to V^*, then the nth shuffle power $e^* \sqcup \ldots \sqcup e^*$ is equal to $n!e^* \otimes \ldots \otimes e^*$.

Prove that the Itô signature of the Brownian motion is not a weakly geometric rough path.

In example 3.2, we saw that a multiplicative functional of degree 1 is simply a path described through its increments. If $p < 2$ and the path has finite p-variation, then this functional is a p-rough path. Furthermore, since $G^{(1)} = V$, it follows that for $p < 2$, the weakly geometric p-rough paths are precisely the functionals of degree 1 and finite p-variation with $p < 2$.

In example 3.3, we looked at a multiplicative functional with values in $T^{(2)}(V)$. Let us now take $V = \mathbb{R}^d$ ($d \geq 2$). We can thus think of generic elements of V as columns $(x^i)_{i=1\ldots d}$ and of 2-tensors as arrays $(x^{ij})_{i,j=1\ldots d}$. Suppose we have a path x_u of finite 1-variation in $\mathbb{R}^d$. We define

$$X_{s,t}^i = x_t^i - x_s^i \quad \text{and} \quad X_{s,t}^{ij} = \iint_{s<u_1<u_2<t} dx_{u_1}^i \, dx_{u_2}^j.$$

Then $X_{s,t} = \left(1, (X_{s,t}^i)_{i=1}^d, (X_{s,t}^{ij})_{i,j=1}^d\right)$ is a multiplicative functional in the space $T^{(2)}(\mathbb{R}^d)$ and has finite p-variation for $1 \leq p < 3$. This functional, which belongs to $G\Omega_p(\mathbb{R}^d)$, is called the canonical extension of the path x_u.

Let us look more carefully at the 2-tensor $(X_{s,t}^{ij})_{i,j=1}^d$. We can express this tensor as the sum of its symmetric and anti-symmetric parts:

$$X_{s,t}^{ij} = \frac{1}{2}(x_t^i - x_s^i)(x_t^j - x_s^j) + A_{s,t}^{ij}, \tag{3.6}$$

where $A_{s,t}^{ij} = \frac{1}{2} \iint_{s<u_1<u_2<t} dx_{u_1}^i \, dx_{u_2}^j - dx_{u_1}^j \, dx_{u_2}^i$ for $i, j = 1, 2, \ldots, d$. The anti-symmetric part of this 2-tensor has a very explicit geometric interpretation. In fact for any two indices i and j, the pair (x_u^i, x_u^j) defines a directed planar curve. We can then add the chord from the point (x_t^i, x_t^j) to (x_s^i, x_s^j) to obtain a closed curve. The integral $A_{s,t}^{ij}$ is then the area enclosed by this closed curve, provided orientation and multiplicity are taken into account. Thus integrating the winding number of the curve over the plane gives $A_{s,t}^{ij}$.

In this particular case, we can identify $G^{(2)}$ in very concrete terms. Indeed, an element of the group $G^{(2)}$ is given by an increment, which belongs to V, and an area, that is, an anti-symmetric 2-tensor on V.

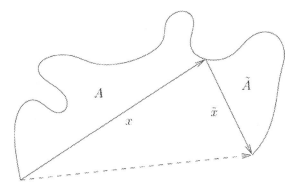

Fig. 3.1. Group multiplication in $G^{(2)}$

For (x, A), $(\tilde{x}, \tilde{A})$ in $G^{(2)}$, the group multiplication is given by:

$$(x, A) \cdot (\tilde{x}, \tilde{A}) = (x + \tilde{x}, \frac{1}{2}(x \wedge \tilde{x}) + A + \tilde{A}),$$

where $\frac{1}{2}(x \wedge \tilde{x})$ is the area of the triangle with sides x and $\tilde{x}$ (Fig. 3.1). This group is embedded into $T^{(2)}(\mathbb{R}^d)$ as follows:

$$(x^i, A^{ij}) \longrightarrow \left(1, x^i, \frac{1}{2}x^i x^j + A^{ij}\right).$$

Hence, extending a path x_t to a geometric rough path involves assigning an area $A_{s,t}$ to each time increment (s, t) so that the identity (3.6) holds.

3.3 The Brownian Rough Path

We are now going to consider some probabilistic examples, and in particular we are going to look in more detail at the Brownian rough path.

3.3.1 The Itô Multiplicative Functional

Let $V = \mathbb{R}^d$ and suppose that $(B_t)_{t \geq 0}$ is an $\mathbb{R}^d$-valued Brownian motion. For each time t, we write $B_t = (B_t^1, \ldots, B_t^d)$ the decomposition of B_t on the canonical basis $(e_1, \ldots, e_d)$ of $\mathbb{R}^d$. We define the following multiplicative functional with values in $T^2(\mathbb{R}^d)$:

$$I_{s,t} = \left(1, I_{s,t}^1, I_{s,t}^2\right)$$

where

$$I_{s,t}^1 = \sum_{i=1}^d I_{s,t}^{1,i} e_i = \sum_{i=1}^d (B_t^i - B_s^i)e_i = B_t - B_s$$

and

$$I_{s,t}^2 = \sum_{i,j=1}^d I_{s,t}^{2,ij} e_i \otimes e_j = \sum_{i,j=1}^d \iint_{s<u_1<u_2<t} dB_{u_1}^i \, dB_{u_2}^j \, e_i \otimes e_j.$$

Hence, $I_{s,t}^1$ is the increment of the Brownian motion between s and t, while $I_{s,t}^2$ is the second iterated Itô integral on $[s,t]$. We can split $I_{s,t}^2$ into its symmetric and antisymmetric components:

$$\begin{aligned}
I_{s,t}^2 &= \sum_{1\le i\le j\le d} \frac{I_{s,t}^{2,ij} + I_{s,t}^{2,ji}}{2} (e_i \otimes e_j + e_j \otimes e_i) \\
&+ \sum_{1\le i<j\le d} \frac{I_{s,t}^{2,ij} - I_{s,t}^{2,ji}}{2} (e_i \otimes e_j - e_j \otimes e_i) \\
&= \frac{1}{2} \sum_{1\le i\le j\le d} \left[(B_t^i - B_s^i)(B_t^j - B_s^j) + \delta_{ij}(t-s) \right] (e_i \otimes e_j + e_j \otimes e_i) \\
&+ \sum_{1\le i<j\le d} A_{s,t}^{ij}(e_i \otimes e_j - e_j \otimes e_i)
\end{aligned}$$

where $A_{s,t}^{ij} = \frac{1}{2} \iint_{s<u_1<u_2<t} dB_{u_1}^i \, dB_{u_2}^j - dB_{u_1}^j \, dB_{u_2}^i$, that is, $A_{s,t}$ is the Lévy area of $(B_t)_{t\ge0}$.

Exercise 3.16. Show that for every $\varepsilon > 0$,

$$\sup_{\mathcal{D}\subset[0,t]} \left\| \sum_{\mathcal{D}} \left[\left| B_{t_{i+1}} - B_{t_i} \right|^{2+2\varepsilon} + \left| A_{t_i,t_{i+1}} \right|^{1+\varepsilon} \right] \right\|^{\frac{1}{2}}$$

is a finite random variable with a Gaussian tail.

Note that $\delta_{ij}(t-s)$ also has finite $(1+\varepsilon)$-variation. Using this result, we have the following corollary.

Corollary 3.17. *The Itô 2-multiplicative functional $I_{s,t}$ is a p-rough path for any $2 < p < 3$.*

We note however that it is not a geometric rough path because the symmetric part of its second iterated integral differs (by $\delta_{ij}(t-s)$) from the value it would have if constructed as a limit of smooth paths (see also Exercise 3.15).

3.3.2 The Stratonovich Multiplicative Functional

We have just seen that the Itô multiplicative functional is a p-rough path $(2 < p < 3)$ but not a geometric rough path. The situation is different if we consider the Stratonovich multiplicative functional.

Paul Lévy proved that if one considers the dyadic piecewise linear approximations $x_t(n)$ to a Brownian motion $(B_t)_{t \in [0,T]}$, that is, for each $n \geq 0$ the path $x(n)$ which is continuous on $[0,T]$, linear off the dyadic points $\frac{k}{2^n}$ and such that $x(n)_{\frac{k}{2^n}} = B_{\frac{k}{2^n}}$, then the area tensor $A(n)$ associated with $x(n)$, characterised by:

$$A(n)^{ij}_{s,t} = \frac{1}{2} \iint\limits_{s < u_1 < u_2 < t} dx(n)^i_{u_1} \, dx(n)^j_{u_2} - dx(n)^j_{u_1} \, dx(n)^i_{u_2} \quad \forall i,j = 1 \ldots d,$$

converges as n tends to infinity to $A_{s,t}$, the Lévy area of the Brownian motion. Furthermore, it can be proved [11, 79] that if we consider for each $n \geq 0$ the canonical extension $S(n)$ of $x(n)$ to a multiplicative functional of degree 2, namely

$$S(n)_{s,t} = \left(1, \int_s^t dx_u(n), \iint\limits_{s < u_1 < u_2 < t} dx(n)_{u_1} \otimes dx(n)_{u_2} \right),$$

then $(S(n))_{n \in \mathbb{N}}$ is a Cauchy sequence in the p-variation topology for all $p > 2$. Moreover, the limit is given by:

$$S_{s,t} = \left(1, B_t - B_s, \frac{1}{2}(B_t - B_s)^{\otimes 2} + A_{s,t} \right)$$

$$= \left(1, B_t - B_s, \iint\limits_{s < u_1 < u_2 < t} \circ dB_{u_1} \otimes \circ dB_{u_2} \right).$$

We recognise here the truncated Stratonovich signature of $(B_t)_{t \geq 0}$. $S_{s,t}$ is called the canonical Brownian rough path. This is indeed a geometric p-rough path $(2 < p < 3)$, as was proved in [79].

Observe that the canonical Brownian rough path is constructed as an almost sure limit and hence, there is a set of measure zero on which this rough path is not defined. Let us now consider a linear stochastic differential equation driven by Brownian motion, of the form

$$dY_t = A(Y_t) \circ dB_t \qquad Y_0 = y,$$

where A is some linear operator. We will prove in Chap. 4 that

$$Y_t = \sum_{n=0}^{\infty} A^{\otimes n} S_{0,t}^n \, y$$

is a solution to this differential equation, where $S_{s,t}^n$ is the nth component of the Stratonovich signature of the Brownian motion. By the extension theorem (Theorem 3.7) we know that given $S_{s,t}^1$ and $S_{s,t}^2$, we can construct all the terms

$S_{s,t}^m$, for $m > 2$. What is remarkable here is that probability only comes in through the construction of the Lévy area. Hence we only have one set of measure zero where the solution is not defined, and this is the same for all suitable operators A. Furthermore, since the process of solving this equation is completely deterministic, once $S_{s,t}^2$ is given, there is no requirement for predictability. In other words, once we have chosen a version of the Lévy area of our Brownian motion, it is possible to solve *simultaneously*, that is, outside a single negligible set, all linear differential equations driven by this Brownian motion.

From the earlier discussion we also get a proof for the Wong-Zakai theorem for linear stochastic differential equations.

Theorem 3.18. *If the paths* $x(n), n \geq 0$ *are the dyadic piecewise linear approximations to a Brownian motion* $(B_t)_{t \in [0,T]}$ *and for each* $n \geq 0$, $y(n)$ *solves the differential equation*

$$dy(n)_t = A(y(n)_t) \, dx(n)_t \qquad y(n)_0 = a,$$

then the sequence $(y(n))_{n \geq 0}$ *converges to* Y, *where*

$$dY_t = A(Y_t) \circ dB_t \qquad Y_0 = y.$$

3.3.3 New Noise Sources

In the classical setting of stochastic differential equations driven by Brownian motion, the techniques described earlier help us to refine our understanding of Itô equations. They can also be extended to equations driven by semi-martingales.

In [17], it is proved that if x_t is a path of finite p-variation in a Banach space V, then it is always possible to extend it to a rough path, and then solve differential equations. Each different extension produces different solutions. What is more interesting is the fact that the method used to construct the canonical Brownian rough path, that is, dyadic piecewise linear approximations, works in more diverse situations. In fact this approach has been used to construct the canonical rough paths associated with Free Brownian motion [26, 82], reversible diffusions on $\mathbb{R}^n$ with finite p-variation for $p < 4$ ([21]), Fractional Brownian motion with Hurst index greater than $\frac{1}{4}$ ([29]), Wiener processes in Banach spaces [51], and diffusions on embedded fractals [42]. Therefore we have a method of solving differential equations in settings which are very much different from the usual semi-martingale setting.

Non-Standard Brownian Motion

Brownian motion is widely used in applications to model noise. A typical framework for such applications is the following. Let $(X^\varepsilon)_{\varepsilon > 0}$ be a family of processes that are piecewise smooth on a fine scale. It is then possible to solve differential equations driven by X^ε. Let Y be defined by:

$$dY_t^\varepsilon = f\left(Y_t^\varepsilon\right) dX_t^\varepsilon \qquad Y_0^\varepsilon = y.$$

Now, assume that, by a homogenisation argument or for any other reason, the processes X^ε converge in distribution to a Brownian motion $(W_t)_{t \geq 0}$. Let Y be the solution of the Stratonovich stochastic differential equation

$$dY_t = f\left(Y_t\right) \circ dW_t \qquad Y_0 = y.$$

It seems reasonable then to expect that the processes $(Y^\varepsilon, X^\varepsilon)$ converge in distribution to (Y, W). However, this can be a mistake and a serious source of modelling error. Indeed, one should pay attention to the convergence of the family $(X^\varepsilon)_{\varepsilon > 0}$ as a family of random variables with values in the space of *geometric rough paths*. In general, the limit will only *look like* Brownian motion: the limit will be a rough path over Brownian motion, but, in general, not the canonical Brownian rough path. Let us illustrate this with an example.

Delayed Brownian Motion

Suppose that we have a simple sound source, which causes a transverse displacement $(Y_t)_{t \geq 0}$ of a window some distance away.

We can model the sound source using white noise, and then the behaviour of the displacement of the window is described by the differential equation

$$dY_t = f_{r,\theta}\left(Y_t\right) \circ dW_t \qquad Y_0 = 0,$$

where r is the distance of the source from the window, and θ is the angle shown in Fig. 3.2. However, the displacement is also affected by sound taking

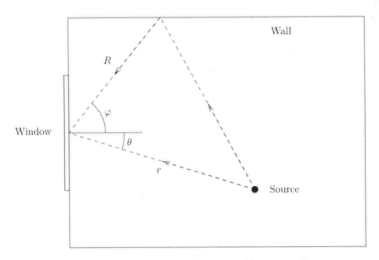

Fig. 3.2. Displacement of a window caused by a sound source

a second route to the window, namely the sound reflecting off the wall. This displacement can be modelled using the equation

$$dY_t = f_{R,\phi}(Y_t) \circ dW_t \qquad Y_0 = 0$$

The first thing that comes to mind is to use the following model for the total displacement:

$$dY_t = (f_{r,\theta}(Y_t) + f_{R,\phi}(Y_t)) \circ dW_t \qquad Y_0 = 0.$$

However, this is not a good model, because if we look carefully, we see that the two sounds take slightly different times to arrive. In fact, the model should rather be

$$dY_t = f_{r,\theta}(Y_t) \circ dW_t + f_{R,\phi}(Y_t) \circ dW_{t-\varepsilon} \qquad Y_0 = 0.$$

Assuming that $f_{r,\theta}$ and $f_{R,\phi}$ are linear, we can use the Wong-Zakai approach to make sense of Y_t after defining a p-rough path associated with the path $t \longrightarrow (W_t, W_{t-\varepsilon})$. Letting ε tend to 0, we get convergence to (W_t, W_t), which is a one-dimensional process, in the sense that its increments take their values in a line. Hence one would expect this process to have zero area. The surprising fact is that this limit has non-zero area.

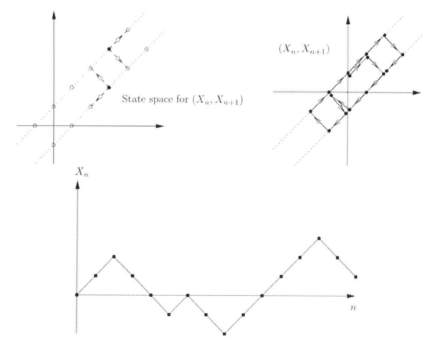

Fig. 3.3. Whatever the walk (X_n) does, the area of the process (X_n, X_{n+1}) is non-increasing.

To understand this, let (X_n) be a simple random walk on $\mathbb{Z}$, that is, a process such that $X_{n+1} = X_n \pm 1$ with equal probability. Consider the process (X_n, X_{n+1}) and extend this to a piecewise linear path. If we compute the increment and area of this path, we can then look at the associated geometric rough path. The point is that, whatever the walk X does, the area of (X_n, X_{n+1}) is monotonic, more precisely non-increasing (see Fig. 3.3).

Going back to our differential equation, it can be shown (cf. [44]) that the limit as ε tends to 0 of $t \mapsto (W_t, W_{t-\varepsilon})$ has a non-trivial linearly growing area. The correct equation for the displacement of the window should thus be

$$dY_t = (f_{r,\theta}(Y_t) + f_{R,\phi}(Y_t)) \circ dW_t + \lambda [f_{r,\theta}, f_{R,\phi}] \, dt \qquad Y_0 = 0.$$

From the earlier discussion we can see that $\lim_{\varepsilon \to 0} (W_., W_{.-\varepsilon})$ looks like $(W_., W_.)$, but it is different from this path as a control.

4

Integration Along Rough Paths

In order to define what we mean by a solution of a differential equation driven by a rough path, we need a theory of integration for rough paths. The appropriate concept turns out to be the integral of a one-form along a rough path. We introduce the notion of almost rough path and prove what is after Theorem 3.7 our second fundamental technical tool, namely the fact that every almost rough path determines a unique rough path. Then, we explain a general strategy for defining a functional on a space of rough paths. This strategy involves some important combinatorics and we practice it on the instructive example of linear equations driven by rough paths. Finally, we tackle the non-linear problem of defining the integral of a one-form along a rough path. As a by-product, we get a change of variable formula, that is, we are able to define the image of a rough path by a well-behaved function.

4.1 Almost-Multiplicativity

4.1.1 Almost-Additivity

As a warm-up, let us review the familiar problem of constructing the indefinite integral of a continuous function from the point of view of almost-additivity.

Consider a function $h : [0, T] \longrightarrow \mathbb{R}$ which belongs to $\mathrm{Lip}(\alpha)$ for some $\alpha \in (0, 1]$, that is, assume that h is Hölder continuous with exponent α.

In order to define the mapping $(s, t) \mapsto \int_s^t h(u) \, du$, which we call *additive* by virtue of the Chasles relation, we start by considering for all $(s, t) \in \triangle_T$

$$H(s, t) = h(s)(t - s). \tag{4.1}$$

The function $H : \triangle_T \longrightarrow \mathbb{R}$ thus defined has the very nice property of being *almost-additive*. This means that there exists two real constants $K \geq 0$ and $\theta > 1$ such that, for all $s, u, t \in [0, T]$ with $s < u < t$,

$$|H(s, u) + H(u, t) - H(s, t)| \leq K|t - s|^{\theta}.$$

In the present setting, this inequality holds with $K = \|h\|_{\mathrm{Lip}(\alpha)}$ and $\theta = 1+\alpha$. This property of almost-additivity is the key to the next proposition, which is a simple instance of the main result of this section.

Proposition 4.1. *Let* $h : [0, T] \longrightarrow \mathbb{R}$ *be* $\mathrm{Lip}(\alpha)$ *for some* $\alpha \in (0, 1]$. *Let* $H : \triangle_T \longrightarrow \mathbb{R}$ *be defined by (4.1). Then there exists a function* $\psi : [0, T] \longrightarrow \mathbb{R}$, *unique up to an additive constant, such that*

$$\sup_{0 \leq s < t \leq T} \frac{|\psi(t) - \psi(s) - H(s, t)|}{(t - s)^{1+\alpha}} < +\infty. \tag{4.2}$$

Since $\psi(t) = \int_0^t h(u)\, du$ satisfies (4.2), this proposition says that $(s, t) \mapsto \int_s^t h(u)\, du$ is the unique real-valued additive functional on $\triangle_T$ which is close to H in the sense of (4.2).

Proof. Uniqueness – If ψ and ψ' both satisfy (4.2), then $\rho = \psi - \psi'$ satisfies for all $(s, t) \in \triangle_T$ and all $n \geq 1$ the inequality

$$|\rho(t) - \rho(s)| \leq \sum_{k=0}^{n-1} \left| \rho\left(s + \frac{k+1}{n}(t - s)\right) - \rho\left(s + \frac{k}{n}(t - s)\right) \right|$$
$$\leq 2Cn^{-\alpha}|t - s|,$$

where C is the left hand side of (4.2). Letting n tend to infinity shows that ρ is a constant function.

Existence – Let us choose (s, t) in $\triangle_T$. For every partition $\mathcal{D}$ of $[s, t]$ with $\mathcal{D} = (s = t_0 < t_1 < \ldots < t_r = t)$, let us define

$$\int_{s, \mathcal{D}}^t h(u)\, du = \sum_{i=0}^{r-1} H(t_i, t_{i+1}). \tag{4.3}$$

The by now usual strategy of removing points from $\mathcal{D}$ one after another until only s and t are left, leads to the maximal inequality

$$\left| \int_{s, \mathcal{D}}^t h(u)\, du - H(s, t) \right| \leq 2^{1+\alpha}(\zeta(1 + \alpha) - 1)\|h\|_{\mathrm{Lip}(\alpha)}|t - s|^{1+\alpha}. \tag{4.4}$$

The same kind of arguments we used in the proof of Theorem 3.7 show that the limit as the mesh of $\mathcal{D}$ tends to zero of (4.3) exists. Then it is clear that

$$(s, t) \mapsto \lim_{|\mathcal{D}| \to 0} \int_{s, \mathcal{D}}^t h(u)\, du$$

is an additive mapping, hence of the form $\psi(t) - \psi(s)$ for some $\psi : [0, T] \longrightarrow \mathbb{R}$. Finally, the bound (4.2) follows from the maximal inequality (4.4). $\square$

4.1.2 Almost Rough Paths

Almost rough paths are just the tensor algebra-valued version of almost additive functions. From now on in this chapter, $[0, T]$ is a fixed compact interval and V, W denote Banach spaces. We assume that their tensor powers are endowed with norms which satisfy the usual requirements of symmetry and consistency (see Definition 1.25).

Definition 4.2. *Let $p > 1$ be a real number. Let $\omega : \triangle_T \longrightarrow [0, +\infty)$ be a control. A function $X : \triangle_T \longrightarrow T^{(\lfloor p \rfloor)}(V)$ is called an almost p-rough path if:*

1. it has p-variation controlled by ω, i.e.

$$\|X_{s,t}^i\| \leq \frac{\omega(s,t)^{\frac{i}{p}}}{\beta\left(\frac{i}{p}\right)!} \quad \forall (s,t) \in \triangle_T, \forall i = 0, \ldots, \lfloor p \rfloor,$$

2. it is almost-multiplicative, i.e. there exists $\theta > 1$ such that

$$\|(X_{s,u} \otimes X_{u,t})^i - X_{s,t}^i\| \leq \omega(s,t)^\theta \quad \forall s < u < t \in [0, T], \forall i = 0, \ldots, \lfloor p \rfloor.$$

If we want to be more specific, we say that X is a θ-almost p-rough path controlled by ω.

Let us give some examples of almost rough paths. First of all, if $h : [0, T] \longrightarrow \mathbb{R}$ is $\mathrm{Lip}(\alpha)$ for some $\alpha \in (0, 1]$, then $X_{s,t} = (1, h(s)(t - s))$ defines a $(1 + \alpha)$-almost 1-rough path.

Next, consider $X \in \mathcal{V}^p([0, T], V)$ and $Y \in \mathcal{V}^q([0, T], \mathbf{L}(V, W))$ with $\frac{1}{p} + \frac{1}{q} > 1$ and $p < 2$. Then $Z_{s,t} = (1, Y_s(X_t - X_s))$ is a $\left(\frac{1}{p} + \frac{1}{q}\right)$-almost p-rough path.

Finally, let $X : \triangle_T \longrightarrow T^{(n)}(V)$ be a p-rough path with $\lfloor p \rfloor \leq n$. Then $\widetilde{X}_{s,t} = (1, X_{s,t}^1, \ldots, X_{s,t}^n, 0)$ is a $\frac{n+1}{p}$-almost p-rough path in $T^{(n+1)}(V)$.

Of course, the second example is related to the existence of the Young integral $\int Y \, dX$ and the third example to the existence and uniqueness of the multiplicative extension of X to $T^{(n+1)}(V)$. In both cases, we proved the existence of a new object by taking products (or sums) corresponding to arbitrary partitions, establishing a maximal inequality and using it to prove the convergence of the products as the mesh of the partitions tends to zero. We use this strategy for the last time in the following result, which in some sense encapsulates it.

Theorem 4.3 (Almost Rough Paths). *Let $p \geq 1$ and $\theta > 1$ be real numbers. Let $\omega : \triangle_T \longrightarrow [0, +\infty)$ be a control. Let $X : \triangle_T \longrightarrow T^{(\lfloor p \rfloor)}(V)$ be a θ-almost p-rough path controlled by ω. Then there exists a unique p-rough path $\widehat{X} : \triangle_T \longrightarrow T^{(\lfloor p \rfloor)}(V)$ such that*

$$\sup_{\substack{0 \leq s < t \leq T \\ i=0,\ldots,\lfloor p \rfloor}} \frac{\|\widehat{X}_{s,t}^i - X_{s,t}^i\|}{\omega(s,t)^\theta} < +\infty. \tag{4.5}$$

Moreover, there exists a constant K which depends only on p, θ and $\omega(0, T)$ such that the supremum above is smaller than K and the p-variation of $\widehat{X}$ is controlled by $K\omega$.

Proof. Uniqueness – Assume that $\widehat{X}$ and $\widetilde{X}$ both satisfy (4.5). Let us prove by induction on $j \in \{0, \ldots, \lfloor p \rfloor\}$ that $\pi_j(\widehat{X}) = \pi_j(\widetilde{X})$, that is, $\widehat{X}$ and $\widetilde{X}$ agree up to level j.

For $j = 0$, this is true. Now assume that $\pi_{j-1}(\widehat{X}) = \pi_{j-1}(\widetilde{X})$ for some j between 1 and $\lfloor p \rfloor$. Then a simple computation stated earlier as Lemma 3.4 shows that $\delta = \widehat{X}^j - \widetilde{X}^j : \triangle_T \longrightarrow V^{\otimes j}$ is an additive functional, that is, the relation $\delta(s, u) + \delta(u, t) = \delta(s, t)$ holds for all $0 \leq s < u < t \leq T$. Thus there exists $\rho : [0, T] \longrightarrow V^{\otimes j}$ such that $\widehat{X}^j_{s,t} - \widetilde{X}^j_{s,t} = \rho(t) - \rho(s)$ for all $(s, t) \in \triangle_T$. Now, (4.5) implies that ρ is a constant function, so that $\widehat{X}$ and $\widetilde{X}$ agree up to level j.

Existence – The proof of the existence of $\widehat{X}$ is very similar to and only slightly more complicated than the proof of Theorem 3.7.

We use an induction argument. For each $j \in \{1, \ldots, \lfloor p \rfloor\}$, we prove the following statement.

Upgrading at level j – Let $Y : \triangle_T \longrightarrow T^{(\lfloor p \rfloor)}(V)$ be a θ-almost p-rough path controlled by ω. Assume that $\pi_{j-1}(Y)$ is multiplicative. Then there exists a θ-almost p-rough path $\widetilde{Y} : \triangle_T \longrightarrow T^{(\lfloor p \rfloor)}(V)$ controlled by $K_j\omega$ such that
(U1) $\widetilde{Y}$ is controlled in p-variation by $K_j\omega$,
(U2) $\|Y^i_{s,t} - \widetilde{Y}^i_{s,t}\| \leq K_j\omega(s, t)^\theta$ for all $(s, t) \in \triangle_T$, all $i \in \{1, \ldots, \lfloor p \rfloor\}$,
(U3) $\pi_j(\widetilde{Y})$ is multiplicative.
Here, K_j is a constant which depends only on j, θ, p and $\omega(0, T)$.

Since $\pi_0(X)$ is certainly multiplicative, $\lfloor p \rfloor$ applications of this result produce a genuine p-rough path which satisfies (4.5) and whose p-variation is controlled by ω times a constant which depends only on p, θ and $\omega(0, T)$.

The proof of the upgrading statement for fixed j is divided in two steps. The first one aims at defining $\widetilde{Y}$. Choose $(s, t) \in \triangle_T$. For each partition $\mathcal{D} = (s = t_0 < t_1 < \ldots < t_r = t)$ of $[s, t]$, set

$$\widetilde{Y}^{\mathcal{D}}_{s,t} = Y_{t_0,t_1} \otimes Y_{t_1,t_2} \otimes \ldots \otimes Y_{t_{r-1},t_r}.$$

Notice that, since $\pi_{j-1}(Y)$ is multiplicative, $\widetilde{Y}^{\mathcal{D}}_{s,t}$ agrees with $Y_{s,t}$ up to level $j - 1$. Then the usual procedure of progressive coarsening of the partition $\mathcal{D}$ leads to the familiar maximal inequality

$$\|(\widetilde{Y}^{\mathcal{D}}_{s,t} - Y_{s,t})^j\| \leq 2^\theta (\zeta(\theta) - 1)\omega(s, t)^\theta. \tag{4.6}$$

Now choose $\varepsilon > 0$ and take $\mathcal{D}$ and $\mathcal{D}'$ two partitions of $[s, t]$ with mesh smaller than ε. Let $\mathcal{D}''$ be a partition of $[s, t]$ which is finer than both $\mathcal{D}$ and $\mathcal{D}'$.

For each $k = 0, \ldots, r-1$, where $r+1$ is the number of points of $\mathcal{D}$, let $\mathcal{D}''_j$ be the partition of $[t_j, t_{j+1}]$ induced by $\mathcal{D}''$. Then $\widetilde{Y}^{\mathcal{D}''}_{s,t} - \widetilde{Y}^{\mathcal{D}}_{s,t}$ is equal to

$$\sum_{k=0}^{r-1} \widetilde{Y}^{\mathcal{D}''_0}_{t_0,t_1} \otimes \ldots \otimes \widetilde{Y}^{\mathcal{D}''_{k-1}}_{t_{k-1},t_k} \otimes \left(\widetilde{Y}^{\mathcal{D}''_k}_{t_k,t_{k+1}} - Y_{t_k,t_{k+1}} \right) \otimes Y_{t_{k+1},t_{k+2}} \otimes \ldots \otimes Y_{t_{r-1},t_r}.$$

If we truncate this expression at level j, there remains only terms of degree exactly j, according to the remark made a few lines above. For the same reason, in each term of the sum, the factor between the brackets has only its term of degree j which is non-zero. Hence, in each term of the sum again, a non-zero term of degree j can arise only by picking the term of degree 0 in each factor, except for the factor in the brackets, for which one has to pick the term of degree j. We have thus proved that

$$\left(\widetilde{Y}^{\mathcal{D}''}_{s,t} - \widetilde{Y}^{\mathcal{D}}_{s,t} \right)^j = \sum_{k=0}^{r-1} \left(\widetilde{Y}^{\mathcal{D}''_k}_{t_k,t_{k+1}} - Y_{t_k,t_{k+1}} \right)^j.$$

Now the maximal inequality (4.6) allows us to estimate the right-hand side of this equality and we get

$$\left\| \left(\widetilde{Y}^{\mathcal{D}''}_{s,t} - \widetilde{Y}^{\mathcal{D}}_{s,t} \right)^j \right\| \le 2^\theta (\zeta(\theta) - 1) \sum_{k=0}^{r-1} \omega(t_k, t_{k+1})^\theta$$

$$\le 2^\theta (\zeta(\theta) - 1) \omega(s,t) \max_{k=0 \ldots r-1} \omega(t_k, t_{k+1})^{\theta-1}$$

$$\le 2^\theta (\zeta(\theta) - 1) \omega(s,t) \sup_{\substack{0 \le u < v \le T \\ |v-u| < \varepsilon}} \omega(u,v)^{\theta-1}.$$

By the uniform continuity of ω, the last expression tends to zero with ε. This estimation and the similar one for $\mathcal{D}'$ and $\mathcal{D}''$ imply immediately the existence of the limit of $\left(\widetilde{Y}^{\mathcal{D}}_{s,t} \right)^j$ as the mesh of $\mathcal{D}$ tends to zero.

We can now define $\widetilde{Y}_{s,t}$ for each $(s,t) \in \triangle_T$ as follows. For each $i \in \{0, \ldots, \lfloor p \rfloor\}$, set

$$\widetilde{Y}^i_{s,t} = \begin{cases} Y^i_{s,t} & \text{if } i \neq j, \\ \lim_{|\mathcal{D}| \to 0} \left(\widetilde{Y}^{\mathcal{D}}_{s,t} \right)^j & \text{if } i = j. \end{cases}$$

The multiplicativity of $\pi_j(\widetilde{Y})$ is a direct consequence of its definition and the multiplicativity of $\pi_{j-1}(Y)$. Now the p-variation control (U1) of $\widetilde{Y}$ and the bound (U2) on $\|Y^i_{s,t} - \widetilde{Y}^i_{s,t}\|$ both follow easily from the maximal inequality (4.6), with some temporary value of the constant K_j.

The second step of the proof of the upgrading result consists in proving that $\widetilde{Y}$ is still θ-almost multiplicative. This is in fact a consequence of the conditions (U1) and (U2), regardless of how $\widetilde{Y}$ has been defined. This explains why we have kept the redundant condition (U1).

Let $s < u < t$ be three times in $[0, T]$. Since $\pi_j(\widetilde{Y})$ is multiplicative, the only non-zero terms of $\widetilde{Y}_{s,u} \otimes \widetilde{Y}_{u,t} - \widetilde{Y}_{s,t}$ are of degree greater than j. Fix $n > j$. Let us define the functional $R : \triangle_T \longrightarrow T^{(\lfloor p \rfloor)}(V)$ by $R_{s,t} = \widetilde{Y}_{s,t} - Y_{s,t}$. We have

$$
\begin{aligned}
\left(\widetilde{Y}_{s,u} \otimes \widetilde{Y}_{u,t} - \widetilde{Y}_{s,t} \right)^n &= \sum_{k=0}^{n} \widetilde{Y}_{s,u}^k \otimes \widetilde{Y}_{u,t}^{n-k} - \widetilde{Y}_{s,t}^n \\
&= \sum_{k=0}^{n} (Y_{s,u}^k + R_{s,u}^k) \otimes (Y_{u,t}^{n-k} + R_{u,t}^{n-k}) - (Y_{s,t}^n + R_{s,t}^n) \\
&= \left[(Y_{s,u} \otimes Y_{u,t})^n - Y_{s,t}^n \right] + \left[(R_{s,u} \otimes R_{u,t})^n - R_{s,t}^n \right] \\
&\quad + \sum_{k=0}^{n} \left(Y_{s,u}^k \otimes R_{u,t}^{n-k} + R_{s,u}^k \otimes Y_{u,t}^{n-k} \right).
\end{aligned}
$$

The last expression splits into three terms. Since Y is θ-almost multiplicative controlled by ω, the norm of the first term is smaller than $\omega(s,t)^\theta$. By the property (U2), the norm of the second term is smaller than $K_j \omega(s,t)^\theta + \lfloor p \rfloor K_j^2 \omega(s,t)^{2\theta}$. Finally, by (U2) again and the fact that Y is controlled by ω, the norm of the third term is smaller than $\frac{2}{\beta} K_j \omega(s,t)^\theta \max(1, \omega(s,t))$. Finally, the inequality

$$
\| \widetilde{Y}_{s,u} \otimes \widetilde{Y}_{u,t} - \widetilde{Y}_{s,t} \| \leq C(\theta, p, \omega(0,T)) K_j \omega(s,t)^\theta
$$

holds for some explicit function C.

Let us replace K_j by $\max(K_j, C(\theta, p, \omega(0,T)) K_j)$. Then (U1) and (U2) continue to hold and we have just proved that $\widetilde{Y}$ is a θ-almost p-rough path controlled by $K_j \omega$. $\square$

Let us quote without proof a result of continuity of the map which to an almost rough path associates a rough path. The following statement is a simplified version of Theorem 3.2.2 of [67].

Theorem 4.4. *Let $p \geq 1$ and $\theta > 1$ be real numbers. Let $\omega : \triangle_T \longrightarrow [0, +\infty)$ be a control. Let $X, Y : \triangle_T \longrightarrow T^{(\lfloor p \rfloor)}(V)$ be two θ-almost p-rough paths controlled by ω. Assume that there exists $\varepsilon > 0$ such that*

$$
\| X_{s,t}^i - Y_{s,t}^i \| \leq \varepsilon \omega(s,t)^{\frac{i}{p}} \quad \forall (s,t) \in \triangle_T, \forall i = 0, \ldots, \lfloor p \rfloor.
$$

Let $\widehat{X}$ and $\widehat{Y}$ be the p-rough paths associated to X and Y. Then there exists a constant $B(\varepsilon, p, \theta, \omega(0,T))$ such that $\lim_{\varepsilon \to 0} B = 0$ and

$$
\| \widehat{X}_{s,t}^i - \widehat{Y}_{s,t}^i \| \leq B \omega(s,t)^{\frac{i}{p}} \quad \forall (s,t) \in \triangle_T, \forall i = 0, \ldots, \lfloor p \rfloor.
$$

We have mentioned at the beginning of this paragraph, when looking at examples of almost-multiplicative rough paths, that Theorem 4.3 summarises several of the results proved earlier in these notes. We are going to apply it with its full strength to construct the integral of a one-form along a rough path. However, although Theorem 4.3 captures an important part of the analysis involved in this forthcoming definition, there is a more algebraic and combinatorial aspect of it which we yet have to discuss, and we choose to do this first on the very special problem of linear equations driven by rough paths.

4.2 Linear Differential Equations Driven by Rough Paths

Let us state at once the main theorem that we are going to prove in this section. Recall that $\Omega_p(V)$ denotes the space of p-rough paths in V and $G\Omega_p(V)$ the space of geometric rough paths, defined as always on the interval $[0, T]$.

Theorem 4.5 (Linear Equations). *Let V and W be two Banach spaces. Let A be an element of $\mathbf{L}(V, \mathbf{L}(W, W))$. Let $I_A : \Omega_1(V) \times W \longrightarrow \Omega_1(W)$ be the Itô map which sends (X, ξ) to the solution Y of the equation*

$$dY_t = AY_t \, dX_t = [A(dX_t)](Y_t) \,, \quad Y_0 = \xi.$$

Then, for every real number $p \geq 1$, I_A extends uniquely to a continuous map $G\Omega_p(V) \times W \longrightarrow G\Omega_p(W)$.

Since we are not yet able to define what it means for a rough path to be a solution of a differential equation driven by another rough path, this theorem is certainly the best we could expect. The uniqueness part of the statement is granted by the fact that $\Omega_1(V)$ is dense in $G\Omega_p(V)$, by definition of $G\Omega_p(V)$. Hence, in order to prove the theorem, all we need to do is to define a continuous map from $\Omega_p(V) \times W$ to $\Omega_p(W)$ which extends I_A. Let us describe a general strategy for doing this.

Suppose we have a functional of some class of smooth paths in mind and we want to extend it to a space of rough paths. The first thing to do is to focus on the case of smooth paths and to express for these paths the functional in terms of their iterated integrals. At a heuristical level (and at a logical, if not computational, level for paths of bounded variation), this should always be possible since the signature characterises the path. The expression that we get at this stage may be fairly complicated and we must simplify it by using any legitimate operation on iterated integrals. Typically, we need to shuffle some integrals. Ideally, this will result in an expression of our functional as a linear function of the signature of the path. This will for instance be the case in this section on linear equations. In this ideal case, there is an obvious way to extend the functional to a space of rough paths, provided the factorial decay of their high-order iterated integrals is strong enough to ensure that infinite expansions that may have arisen still converge.

In the special case, which is relevant to our problem, where the functional we have in mind takes its values in a space of paths, we expect its extension to take its values in a space of rough paths, and Theorem 4.3 allows us to cheat to some extent by only approximating the functional by a linear function of the signature. This will typically involve Taylor expansions and small error terms neglected. In general, we will end up with a linear function of the signature of a path which produces non-multiplicative functionals. However, if the approximation is good enough, they will be almost-multiplicative and Theorem 4.3 turns them into rough paths. This will be illustrated by the third section of this chapter.

Let us make a final remark on the appearance of *geometric* rough paths in the statement of Theorem 4.5. In the "ideal" case above, it seems that there is no reason why the linear function of the signature we came up with would be better defined for geometric rough paths rather than general multiplicative functionals. However, in the setting of differential equations, there is. Indeed, it is not at all true that a linear mapping between the tensor algebras over two vector spaces sends multiplicative functionals to multiplicative functionals. In particular, the "solution" to the linear differential equation we would get by blindly applying the strategy described above to a non-geometric rough path would in general not be a multiplicative functional.

On the other hand, as long as we analyse our path-valued functional on smooth paths, we know that we are constructing a path from another path, and this guarantees that we are writing down only multiplicative functionals. The algebraic manipulations which we are allowed to make on iterated integrals are an exact counterpart of the constraints satisfied by the signatures of smooth paths, namely their behaviour with respect to the shuffle product. By definition, these constraints are still satisfied by any geometric rough path, so that the solution of the equation driven by a geometric path will indeed be multiplicative.

After all these comments and warnings, let us turn to the proof of Theorem 4.5. In this proof, we use without further notice the fact that a rough path on a vector space and an element of this vector space determine a classical path in this vector space. If $Y \in \Omega_p(W)$ and $\xi \in W$, we associate to Y and ξ the classical path $t \mapsto \xi + Y_{0,t}^1$, which we denote by Y_t if no confusion is possible.

Proof. Let us choose X in $\Omega_1(V)$ and ξ in W. In Sect. 2.1, we have shown that $Y = I_A(X, \xi)$ can be expressed as follows: for all $(s, t) \in \triangle_T$,

$$Y_t = \sum_{k=0}^{\infty} A^{\otimes k} \int_{s < v_1 < \ldots < v_k < t} dX_{v_1} \otimes \ldots \otimes dX_{v_k} \ Y_s = \sum_{k=0}^{\infty} A^{\otimes k}(X_{s,t}^k) \ Y_s,$$

where by definition, $A^{\otimes n}(x_1 \otimes \ldots \otimes x_n)y = [A(x_n) \circ \ldots \circ A(x_1)](y)$. In order to define Y as a rough path, we need to express its iterated integrals in terms of those of X. For this, let us assume that X is Lipschitz continuous and parametrised at unit speed, that is, $\|\dot{X}_t\| = 1$ for a.e. $t \in [0, T]$. Under this assumption, the derivative of the mapping $t \mapsto X_{s,t}^k$ is given by $X_{s,t}^{k-1} \otimes \dot{X}_t$. We usually write this relation under the form $dX_{s,t}^k = X_{s,t}^{k-1} \otimes dX_t$. We can compute the derivative of Y:

$$\dot{Y}_t = \sum_{k=1}^{\infty} A^{\otimes k} \left[\left(\int_{s < v_1 < \ldots < v_{k-1} < t} \dot{X}_{v_1} \otimes \ldots \otimes \dot{X}_{v_{k-1}} \, dv_1 \ldots dv_{k-1} \right) \otimes \dot{X}_t \right] Y_s$$

$$= \sum_{k=1}^{\infty} A^{\otimes k} (X_{s,t}^{k-1} \otimes \dot{X}_t) Y_s.$$

The norm of the kth term of this series is bounded by $\frac{\|A\|^n}{(n-1)!}$. Thus, the dominated convergence theorem allows us to permute summation and integration when we write down the iterated integrals of Y:

$$Y_{s,t}^n = \int_{s < u_1 < \ldots < u_n < t} \dot{Y}_{u_1} \otimes \ldots \otimes \dot{Y}_{u_n} \, du_1 \ldots du_n$$

$$= \sum_{k_1, \ldots, k_n = 1}^{\infty} \int_{s < u_1 < \ldots < u_n < t} A^{\otimes k_1} (X_{s,u_1}^{k_1 - 1} \dot{X}_{u_1}) Y_s \ldots A^{\otimes k_n} (X_{s,u_n}^{k_n - 1} \dot{X}_{u_n}) Y_s \, du$$

$$= \sum_{k_1, \ldots, k_n = 1}^{\infty} A^{\otimes (k_1 + \ldots + k_n)} \int_{s < u_1 < \ldots < u_n < t} dX_{s,u_1}^{k_1} \otimes \ldots \otimes dX_{s,u_n}^{k_n} (Y_s)^{\otimes n}. \quad (4.7)$$

In this computation, some tensor products are implicit and du stands for $du_1 \ldots du_n$. The integral in the last expression would not make sense without the assumption that X is Lipschitz continuous. It can however be written in a way that does make sense for arbitrary rough paths. Let us first expand it when X is Lipschitz continuous. It is equal to

$$\int_{\sigma} \dot{X}_{v_{1,1}} \ldots \dot{X}_{v_{1,k_1}} \dot{X}_{v_{2,1}} \ldots \dot{X}_{v_{2,k_2}} \ldots \dot{X}_{v_{n,1}} \ldots \dot{X}_{v_{n,k_n}} \, dv. \quad (4.8)$$

In this integral, tensor products are again implicit and dv stands for the Lebesgue measure on the domain $\sigma \subset \mathbb{R}^{k_1 + \ldots + k_n}$ defined by the inequalities

$$s < v_{1,1} < \ldots < v_{1,k_1} < t,$$
$$s < v_{2,1} < \ldots < v_{2,k_2} < t,$$
$$\vdots$$
$$s < v_{n,1} < \ldots < v_{1,k_n} < t,$$
$$\text{and} \quad s < v_{1,k_1} < \ldots < v_{n,k_n} < t.$$

Would the times corresponding to the successive factors in (4.8) be increasing, it would simply be equal to $X_{s,t}^K$, where we set $K = k_1 + \ldots + k_n$. However, as it is, the integral decomposes as a sum over all possible orderings of $v_{1,1}, \ldots, v_{n,k_n}$.

Remember that an element π of the symmetric group $\mathfrak{S}_K$ acts on $V^{\otimes K}$ by sending $x_1 \otimes \ldots \otimes x_K$ to $x_{\pi(1)} \otimes \ldots \otimes x_{\pi(K)}$. The permutations of $\{1, \ldots, K\}$ which arise on the domain σ are those we call *ordered shuffles*: we define $OS(k_1, \ldots, k_n)$ to be the subset of $\mathfrak{S}_K$ consisting of those π for which

$$\pi(1) < \pi(2) < \ldots < \pi(k_1) \,, \quad \pi(k_1 + 1) < \ldots < \pi(k_1 + k_2) \,, \ldots,$$

$$\pi(K - k_n + 1) < \ldots < \pi(K) \text{ and } \pi(k_1) < \pi(k_2) < \ldots < \pi(k_n).$$

The last condition distinguishes them among ordinary shuffles. We have the following equality:

$$\int_{s < u_1 < \ldots < u_n < t} dX^{k_1}_{s,u_1} \otimes \ldots \otimes dX^{k_n}_{s,u_n} = \sum_{\pi \in OS(k_1, \ldots, k_n)} \pi^{-1} X^K_{s,t}. \tag{4.9}$$

Let us use the shorthand notation $\mathbf{k} = (k_1, \ldots, k_n)$ and $|\mathbf{k}| = k_1 + \ldots + k_n$. For each $(s,t) \in \triangle_T$ and each $n \geq 1$, we have

$$Y^n_{s,t} = \sum_{\mathbf{k} \in (\mathbb{N}^*)^n} A^{\otimes |\mathbf{k}|} \sum_{\pi \in OS(\mathbf{k})} \pi^{-1} X^{|\mathbf{k}|}_{s,t} \, Y^{\otimes n}_s. \tag{4.10}$$

There is some implicit ordering of the factors in this compact expression. Let us emphasise that by $A^{\otimes |\mathbf{k}|}(x_1 \otimes \ldots \otimes x_{|\mathbf{k}|}) y^{\otimes n}$, we mean

$$[A(x_{k_1}) \circ \ldots \circ A(x_1)](y) \otimes [A(x_{k_1 + k_2}) \circ \ldots \circ A(x_{k_1 + 1})](y) \otimes \ldots$$
$$\ldots \otimes [A(x_{|\mathbf{k}|}) \circ \ldots \circ A(x_{|\mathbf{k}| - k_n + 1})](y).$$

We have now got an expression of Y which is linear in the signature of X. In order to control the convergence of the series when X is a rough path, we need to estimate the size of the sets of ordered shuffles. We can do this in a crude way by fixing K and observing that every element of $\bigcup_{|\mathbf{k}| = K} OS(\mathbf{k})$ determines a colouring of $\{1, \ldots, K\}$ in n colours. Since there are less than n^K such colourings, we find $\sum_{|\mathbf{k}| = K} \mathrm{Card}[OS(\mathbf{k})] \leq n^K$. Hence, if X is a p-rough path with p-variation controlled by some control ω, we find

$$\left\| \sum_{|\mathbf{k}| = K} A^{\otimes K} \sum_{\pi \in OS(\mathbf{k})} \pi^{-1} X^K_{s,t} \right\| \leq \|A\|^K n^K \frac{\omega(s,t)^{\frac{K}{p}}}{\beta \left(\frac{K}{p} \right)!}.$$

Hence, the right-hand side of (4.10) makes sense for any p-rough path X.

It defines a map which extends the Itô map and which we also denote by I_A. It is not difficult to check that I_A is continuous in the p-variation topology.

Then, $I_A : \Omega_p(V) \longrightarrow \Omega_p(W)$ sends $\Omega_1(V)$ into $\Omega_1(W) \subset G\Omega_p(W)$. Since $\Omega_1(V)$ is dense in $G\Omega_p(V)$ and $G\Omega_p(W)$ is closed in $\Omega_p(W)$, I_A sends $G\Omega_p(V)$ into $G\Omega_p(W)$. $\qquad \square$

4.3 Integration of a One-form Along a Rough Path

We are now going to apply the ideas discussed so far in this chapter to define a notion of integral along a rough path.

It is important to realise that, if X and Y are two p-rough paths with $p \geq 2$ and f is a map, even the smoothest one, there cannot exist a general sensible definition of something like $\int f(Y)\, dX$. Indeed, as the simple example of $\int_s^t Y_u\, dX_u = \int_{s < u_1 < u_2 < t} dY_{u_1} dX_{u_2}$ illustrates, such a definition would necessarily involve some cross-iterated integrals of X and Y, which are not available in the data of X and Y. We are reformulating the fact that two rough paths X and Y, the first in $\Omega_p(V)$, the second in $\Omega_p(W)$, do not at all determine the pair (X, Y) as an element of $\Omega_p(V \oplus W)$.

On the other hand, if one knows the pair $Z = (X, Y)$ as a rough path, then $\int f(Y)\, dX$ is a special case of $\int \alpha(Z)\, dZ$, where α is a one-form. This is the expression to which we are going to give a meaning.

4.3.1 Construction of an Almost Rough Path

Let $\gamma > p \geq 1$ be real numbers. Let $Z : \triangle_T \longrightarrow T^{(\lfloor p \rfloor)}(V)$ be a p-rough path. Let $\alpha : V \longrightarrow \mathbf{L}(V, W)$ be a Lip$(\gamma - 1)$ function, or one-form as we will call it from now on. Thus, we are given $\alpha^0 = \alpha$ and at least $\lfloor p \rfloor - 1$ auxiliary functions $\alpha^1, \ldots, \alpha^{\lfloor p \rfloor - 1}$, where for each $l = 1, \ldots, \lfloor p \rfloor - 1$,

$$\alpha^l : V \longrightarrow \mathbf{L}(V^{\otimes l}, \mathbf{L}(V, W))$$

and for all $v \in V$, $\alpha^l(v)$ is a *symmetric* l-linear mapping from V to $\mathbf{L}(V, W)$. These functions satisfy the Taylor-like relations described in Definition 1.21. For simplicity, let us assume that there are no other auxiliary functions, that is, $\gamma - 1 \in (\lfloor p \rfloor - 1, \lfloor p \rfloor]$. This does not put any further restriction on α, it only makes the Taylor bounds easier to write down.

Let us first look for an approximation of $\int_s^t \alpha(Z_u)\, dZ_u$, that is, the first degree term of the integral, by using the Taylor expansion of α. We know that, for all $x, y \in V$,

$$\alpha(y) = \alpha(x) + \sum_{l=1}^{\lfloor p \rfloor - 1} \alpha^l(x) \frac{(y - x)^{\otimes l}}{l!} + R_0(x, y)$$

and $\|R_0(x, y)\| \leq \|\alpha\|_{\mathrm{Lip}} \|x - y\|$. In the remarks made after Definition 1.21, we observed that the symmetry of the multilinear forms $\alpha^l(x)$ allows us to replace x and y by the initial and final points of a path Z with bounded variation on an interval $[s, u]$ and $(y - x)^{\otimes l}/l!$ by $Z_{s,u}^l$. From now on, let us assume that Z has finite 1-variation controlled by some ω. Then, for all $(s, u) \in \triangle_T$,

$$\alpha(Z_u) = \sum_{l=0}^{\lfloor p \rfloor - 1} \alpha^l(Z_s) Z_{s,u}^l + R_0(Z_s, Z_u). \tag{4.11}$$

On the other hand, by definition of the iterated integrals, the following relation holds for all $l \geq 0$:

$$\int_s^t Z_{s,u}^l \otimes dZ_u = Z_{s,t}^{l+1}. \tag{4.12}$$

Combining (4.11) and (4.12), we find

$$\int_s^t \alpha(Z_u)\, dZ_u = \sum_{l=0}^{\lfloor p \rfloor - 1} \alpha^l(Z_s) Z_{s,t}^{l+1} + \int_s^t R_0(Z_s, Z_u)\, dZ_u.$$

The order of magnitude of the error term is:

$$\left\| \int_s^t R_0(Z_s, Z_u)\, dZ_u \right\| \simeq \|R_0(Z_s, Z_t)\| \|Z_{s,t}^1\| \simeq \omega(s,t)^{\frac{\gamma}{p}}.$$

Since $\gamma > p$, there is some hope that dropping this error term will leave us with an almost-additive approximation of the first-degree term of the integral. We will prove in a while that this is true. Let us define

$$Y_{s,t}^1 = \sum_{l=0}^{\lfloor p \rfloor - 1} \alpha^l(Z_s) Z_{s,t}^{l+1}. \tag{4.13}$$

We must now define the higher-degree terms of Y in such a way that it becomes an almost rough path. We can try to guess what they should be from (4.13). Let us fix $n \geq 2$ and try to compute $Y_{s,t}^n$.

Before we do this, let us make the following convention. It is sometimes convenient to replace a summation over $l = 0, \ldots, \lfloor p \rfloor - 1$ by a summation over $k = 1, \ldots, \lfloor p \rfloor$. We consistently keep indices called l ranging from 0 to $\lfloor p \rfloor - 1$ and those called k ranging from 1 to $\lfloor p \rfloor$. In other words, $k_* = l_* + 1$.

$$
\begin{aligned}
Y_{s,t}^n &= \int_{s<u_1<\ldots<u_n<t} dY_{s,u_1}^1 \otimes \ldots \otimes dY_{s,u_n}^1 \\
&\simeq \int_{s<u_1<\ldots<u_n<t} \sum_{l_1=0}^{\lfloor p \rfloor - 1} \alpha^{l_1}(Z_s) dZ_{s,u_1}^{l_1+1} \otimes \ldots \otimes \sum_{l_n=0}^{\lfloor p \rfloor - 1} \alpha^{l_n}(Z_s) dZ_{s,u_n}^{l_n+1} \\
&= \sum_{k_1,\ldots,k_n=1}^{\lfloor p \rfloor} \alpha^{k_1-1}(Z_s)\ldots\alpha^{k_n-1}(Z_s) \int_{s<u_1<\ldots<u_n<t} dZ_{s,u_1}^{k_1} \otimes \ldots \otimes dZ_{s,u_n}^{k_n}.
\end{aligned}
$$

We recognise here the same integral as in (4.7) and as in the linear case, we use ordered shuffles to express $Y_{s,t}^n$ in terms of the iterated integrals of Z, according to (4.9):

$$Y_{s,t}^n = \sum_{k_1,\ldots,k_n=1}^{\lfloor p \rfloor} \alpha^{k_1-1}(Z_s) \otimes \ldots \otimes \alpha^{k_n-1}(Z_s) \sum_{\pi \in OS(k_1,\ldots,k_n)} \pi^{-1} Z_{s,t}^{k_1+\ldots+k_n}.$$

This our best guess for a sensible approximation to $\int \alpha(Z)\, dZ$. The main theorem of this chapter should now sound plausible.

4.3.2 Definition of the Integral

Theorem 4.6. *Let $Z : \triangle_T \longrightarrow T^{(\lfloor p \rfloor)}(V)$ be a geometric p-rough path. Let γ be such that $\gamma > p$. Let $\alpha : V \longrightarrow \mathbf{L}(V, W)$ be a $\mathrm{Lip}(\gamma - 1)$ function. Then $Y : \triangle_T \longrightarrow T^{(\lfloor p \rfloor)}(W)$ defined for all $(s, t) \in \triangle_T$ and all $n \geq 1$ by*

$$Y_{s,t}^n = \sum_{k_1,\ldots,k_n=1}^{\lfloor p \rfloor} \alpha^{k_1-1}(Z_s) \otimes \ldots \otimes \alpha^{k_n-1}(Z_s) \sum_{\pi \in OS(k_1,\ldots,k_n)} \pi^{-1} Z_{s,t}^{k_1+\ldots+k_n} \tag{4.14}$$

is an almost p-rough path.

Let us emphasise the fact that the almost-rough path Y depends on α and on all its auxiliary functions. If we start with a one-form α which is defined on a thin subset of V, like the range of Z, then choose auxiliary functions for α and finally extend this collection of functions to an element of $\mathrm{Lip}(V)$ by Whitney's theorem, then we must keep in mind that the pseudo-derivatives of α are in general not determined by α, or at least not completely. Thus, the integral of the one-form α will depend on the particular choice of the extension of α as a Lip function.

Let us now explain the principle of the proof, which is simpler than it seems. It consists in *computing*, rather than estimating, the difference $(Y_{s,u} \otimes Y_{u,t})^n - Y_{s,t}^n$ for each n. We do this first in the case where Z has bounded variation, and then use the fact that an equality between quantities which are continuous in the p-variation topology holds for all geometric p-rough paths if it holds for paths with bounded variation.

In the computation of the defect of multiplicativity, some terms appear with a derivative of α evaluated at Z_s and some with a derivative evaluated at Z_u. It is crucial to express everything in terms of the value at Z_s of the derivatives. The following lemma shows that this can be done in a neat way by using the Taylor expansion. Recall that $R_0, \ldots, R_{\lfloor p \rfloor - 1}$ are the error terms in the Taylor expansions of α and its derivatives.

Lemma 4.7. *For all $x, y, z, v \in V$,*

$$\sum_{l=0}^{\lfloor p \rfloor - 1} \left(\alpha^l(y) - R_l(x, y) \right) \left(\frac{(z-y)^{\otimes l}}{l!} \right)(v) = \sum_{l=0}^{\lfloor p \rfloor - 1} \alpha^l(x) \left(\frac{(z-x)^{\otimes l}}{l!} \right)(v). \tag{4.15}$$

In particular, if Z is Lipschitz continuous, then for a.e. $s < u < t$ in $[0, T]$,

$$\sum_{l=0}^{\lfloor p \rfloor - 1} \left(\alpha^l(Z_u) - R_l(Z_s, Z_u) \right) (Z_{u,t}^l)(\dot{Z}_t) = \sum_{l=0}^{\lfloor p \rfloor - 1} \alpha^l(Z_s)(Z_{s,t}^l)(\dot{Z}_t).$$

Proof. The second part of the statement follows from the first because the derivatives of α are symmetric in their second argument and for all $(s, t) \in \triangle_T$, the symmetric part of $Z_{s,t}^l$ is $(Z_{s,t}^1)^{\otimes l}/l!$.

In order to prove the first statement, we simply write down the Taylor relations given by the definition of α being $\mathrm{Lip}(\gamma - 1)$, between the points x and y. Thus, for each $l = 0, \ldots, \lfloor p \rfloor - 1$,

$$\left(\alpha^l(y) - R_l(x,y) \right) \left(\frac{(z-y)^{\otimes l}}{l!} \right)(v)$$

$$= \sum_{m=l}^{\lfloor p \rfloor - 1} \alpha^m(x) \left(\frac{(z-y)^{\otimes l} \otimes (y-x)^{\otimes(m-l)}}{l!(m-l)!} \right)(v).$$

By summing these relations over $l \in \{0, \ldots, \lfloor p \rfloor - 1\}$, we get

$$\text{l.h.s. of (4.15)} = \sum_{0 \le l \le m \le \lfloor p \rfloor - 1} \alpha^m(x) \left(\frac{(z-y)^{\otimes l} \otimes (y-x)^{\otimes(m-l)}}{l!(m-l)!} \right)(v)$$

$$= \sum_{m=0}^{\lfloor p \rfloor - 1} \alpha^m(x) \left(\sum_{l=0}^{m} \frac{(z-y)^{\otimes l} \otimes (y-x)^{\otimes(m-l)}}{l!(m-l)!} \right)(v).$$

For each m, the projection on the space of symmetric tensors of degree m of the sum inside the brackets is equal to $(z-x)^{\otimes m}/m!$. The symmetry property of the derivatives of α implies that the last expression is equal to the r.h.s. of (4.15). □

This lemma will allow us to prove the next result, which computes the defect of multiplicativity of Y. In the following statement, we do not assume that Z is Lipschitz continuous, only that it is geometric.

Lemma 4.8. *For all $s < u < t$ in $[0, T]$,*

$$Y_{s,u} \otimes Y_{u,t} - Y_{s,t} = Y_{s,u} \otimes N_{s,u,t}, \tag{4.16}$$

where $N_{s,u,t}$ is defined as follows.
For $\varepsilon \in \{0, 1\}$ and $l \in \{0, \ldots, \lfloor p \rfloor - 1\}$, set

$$F_l^\varepsilon (Z_s, Z_u) = \begin{cases} R_l(Z_s, Z_u) & \text{if } \varepsilon = 0, \\ -\alpha^l(Z_s) & \text{if } \varepsilon = 1. \end{cases}$$

Then

$$N_{s,u,t}^j = \sum_{\substack{k_1, \ldots, k_j \in \{1, \ldots, \lfloor p \rfloor\} \\ \varepsilon_1, \ldots, \varepsilon_j \in \{0,1\} \\ \varepsilon_1 \ldots \varepsilon_j = 0}} F_{k_1 - 1}^{\varepsilon_1}(Z_s, Z_u) \ldots F_{k_j - 1}^{\varepsilon_j}(Z_s, Z_u) \sum_{\pi \in OS(k_1, \ldots, k_j)} \pi^{-1} Z_{u,t}^{k_1 + \ldots + k_j}.$$

In the definition of $N_{s,u,t}$, the summation over the ε's allows all combinations of R and α, except the combination where each term is an α.

Proof. Let us start by assuming that Z is Lipschitz continuous. Then we can go back to an earlier expression of $Y_{s,t}^n$ valid only in this case:

$$Y_{s,t}^n = \int_{s<v_1<\ldots<v_n<t} \sum_{l_1=0}^{\lfloor p \rfloor -1} \alpha^{l_1}(Z_s)(Z_{s,v_1}^{l_1})(dZ_{v_1}) \ldots \sum_{l_n=0}^{\lfloor p \rfloor -1} \alpha^{l_n}(Z_s)(Z_{s,v_n}^{l_n})(dZ_{v_n}).$$

Now let us fix a degree n and prove (4.16) for the terms of degree n. A straightforward application of Lemma 4.7 gives us

$$\sum_{i=0}^n Y_{s,u}^i \otimes Y_{u,t}^{n-i} = \sum_{i=0}^n Y_{s,u}^i \otimes N_{s,u,t}^{n-i} + \sum_{i=0}^n \int_{\substack{s<v_1<\ldots<v_i<u \\ u<v_{i+1}<\ldots<v_n<t}}$$
$$\times \sum_{l_1=0}^{\lfloor p \rfloor -1} \alpha^{l_1}(Z_s)(Z_{s,v_1}^{l_1})(dZ_{v_1}) \ldots \sum_{l_n=0}^{\lfloor p \rfloor -1} \alpha^{l_n}(Z_s)(Z_{s,v_n}^{l_n})(dZ_{v_n}).$$

The last term is equal to $Y_{s,t}^n$ and the result is proved. Since (4.16) holds for paths with bounded variation and both sides of the equality are continuous in p-variation topology, the result is still true if we only assume that Z is a geometric p-rough path.   □

Let us bring everything together in the proof of Theorem 4.6.

Proof. Assume that the p-variation of Z is controlled by ω. Choose $s < t < u$ in $[0, T]$ and $j \geq 1$.

In the sum which defines $N_{s,u,t}^j$, each term has at least one factor of the form $R_{l_i-1}(Z_s, Z_u)$, whose norm is smaller than

$$\|Z_{s,u}^1\|^{\gamma - l_i} \leq \omega(s,t)^{\frac{\gamma - l_i}{p}},$$

where i is some integer between 1 and j. On the other hand,

$$\|Z_{u,t}^{l_1+\ldots+l_j}\| \leq \frac{\omega(s,t)^{\frac{l_1+\ldots+l_j}{p}}}{\beta\left(\frac{l_1+\ldots+l_j}{p}\right)!}.$$

So, there exists a constant $C(\|\alpha\|_{\mathrm{Lip}(\gamma-1)}, p, \gamma, \omega(0,T))$ such that

$$\|N_{s,u,t}^j\| \leq C\omega(s,t)^{\frac{\gamma}{p}}.$$

There also exists another constant $C(\|\alpha\|_{\mathrm{Lip}(\gamma-1)}, p, \gamma, \omega(0,T))$ such that the p-variation of Y is controlled by $C\omega$. It follows now easily from Lemma 4.8 that there exists a third constant $C(\|\alpha\|_{\mathrm{Lip}(\gamma-1)}, p, \gamma, \omega(0,T))$ such that

$$\|Y_{s,t}^n - (Y_{s,u} \otimes Y_{u,t})^n\| \leq C\omega(s,t)^{\frac{\gamma}{p}}. \tag{4.17}$$

Since $\gamma > p$, Y is $\frac{\gamma}{p}$-almost multiplicative.   □

Definition 4.9 (Integral of a 1-Form). *Let* $Z : \triangle_T \longrightarrow T^{(\lfloor p \rfloor)}(V)$ *be a geometric p-rough path. Let* $\alpha : V \longrightarrow \mathbf{L}(V, W)$ *be a* $\mathrm{Lip}(\gamma - 1)$ *function for some* $\gamma > p$. *Let* $Y : \triangle_T \longrightarrow T^{(\lfloor p \rfloor)}(W)$ *be the almost p-rough path defined by Theorem 4.6. The unique p-rough path associated to* Y *by Theorem 4.3 is called the integral of* α *along* Z *and it is denoted by* $\int_\bullet^\bullet \alpha(Z_u) \, dZ_u :$ $\triangle_T \longrightarrow T^{(\lfloor p \rfloor)}(W)$.

We use the notation $\int_s^t \alpha(Z_u) dZ_u^n$ *to denote the nth degree term of* $\int_s^t \alpha(Z_u) dZ$.

Let us insist again on the fact that the integral of α along Z depends on all the auxiliary functions $\alpha^1, \ldots, \alpha^{\lfloor p \rfloor - 1}$. Thus, if α is for instance a smooth one-form defined only on the range of Z, the definition of the integral requires a choice of $\alpha^1, \ldots, \alpha^{\lfloor p \rfloor - 1}$ and this choice is in general only partially determined by α itself. However, different choices will lead to different values of the integral.

Let us make a slight improvement in this definition. A careful look at the definition (4.14) of $Y_{s,t}^n$ reveals that some terms are superfluous: all the terms corresponding to $(k_1, \ldots, k_n)$ with $k_1 + \ldots + k_n > \lfloor p \rfloor$ produce a contribution which is dominated by $\omega(s,t)^{(\lfloor p \rfloor + 1)/p}$, where ω is a control of the p-variation of Z. They do not affect the fact that Y is almost-multiplicative nor modify the rough path associated to Y. Let us state this.

Proposition 4.10. *Recall the notation introduced in Theorem 4.6. Let* $\widetilde{Y} : \triangle_T \longrightarrow T^{(\lfloor p \rfloor)}(W)$ *be defined by:*

$$\widetilde{Y}_{s,t}^n = \sum_{\substack{k_1, \ldots, k_n \in \{1, \ldots, \lfloor p \rfloor\} \\ k_1 + \ldots + k_n \leq \lfloor p \rfloor}} \alpha^{k_1 - 1}(Z_s) \otimes \ldots \otimes \alpha^{k_n - 1}(Z_s) \sum_{\pi \in OS(k_1, \ldots, k_n)} \pi^{-1} Z_{s,t}^{k_1 + \ldots + k_n}$$

(4.18)

for all $(s,t) \in \triangle_T$ *and all* $n \geq 1$. *Then* $\widetilde{Y}$ *is an almost p-rough path and the p-rough path associated to* $\widetilde{Y}$ *is* $\int \alpha(Z) \, dZ$.

Proof. Let ω be a control of the p-variation of Z. There exists a constant K such that, for all $(s,t) \in \triangle_T$ and all $i = 1, \ldots, \lfloor p \rfloor$,

$$\|Y_{s,t}^i - \widetilde{Y}_{s,t}^i\| \leq K \omega(s,t)^{\frac{\lfloor p \rfloor + 1}{p}}.$$

Just as in the last part of the proof of Theorem 4.3, this and the fact that Y is almost-multiplicative implies that $\widetilde{Y}$ is an almost rough path.

Since $\widetilde{Y}$ and $\int \alpha(Z) \, dZ$ satisfy the bound (4.5), the rough path associated to $\widetilde{Y}$ is also $\int \alpha(Z) \, dZ$. $\square$

Exercise 4.11. Recall the notation of Theorem 4.6 and Proposition 4.10. Assume that $2 \leq p < 3$. Check that, in this case, $\widetilde{Y}_{s,t} = (1, \widetilde{Y}_{s,t}^1, \widetilde{Y}_{s,t}^2)$ is given by:

$$\widetilde{Y}_{s,t}^1 = \alpha(Z_s)Z_{s,t}^1 + \alpha^1(Z_s)Z_{s,t}^2,$$
$$\widetilde{Y}_{s,t}^2 = (\alpha(Z_s) \otimes \alpha(Z_s))Z_{s,t}^2. \tag{4.19}$$

Prove directly that $\widetilde{Y}$ is an almost p-rough path and observe that you can do it without using the assumption that Z is geometric. Hence, if $2 \le p < 3$, the integral $\int \alpha(Z)\,dZ$ is defined for all $Z \in \Omega_p(V)$.

Let us prove that the mapping $Z \mapsto \int \alpha(Z)\,dZ$ is continuous.

Theorem 4.12. *Recall the notation of Definition 4.9. The mapping $Z \mapsto \int \alpha(Z)\,dZ$ is continuous from $G\Omega_p(V)$ to $G\Omega_p(W)$.*

Moreover, if $\omega : \triangle_T \longrightarrow [0,+\infty)$ is a control, there exists a constant $K(\|\alpha\|_{\mathrm{Lip}}, p, \gamma, \omega(0,T))$ such that for all $Z \in G\Omega_p(V)$ with p-variation controlled by ω,

$$\left\| \int_s^t \alpha(Z)\,dZ^i \right\| \le K\omega(s,t)^{\frac{i}{p}} \quad \forall(s,t) \in \triangle_T, \forall i = 0\dots\lfloor p \rfloor. \tag{4.20}$$

Proof. The map which sends Z to Y defined by (4.14) is continuous in the p-variation topology. By Theorem 4.4, the map which sends Y to $\int \alpha(Z)\,dZ$ is also continuous. So, we have a continuous mapping $\int \alpha : \Omega_p(V) \longrightarrow \Omega_p(W)$.

This mapping agrees with the classical Riemann integral on $\Omega_1(V)$, so that it sends $\Omega_1(V)$ into $\Omega_1(W) \subset G\Omega_p(W)$. Since $\Omega_1(V)$ is dense in $G\Omega_p(V)$ and $G\Omega_p(W)$ is closed in $\Omega_p(W)$, $\int \alpha$ sends $G\Omega_p(V)$ into $G\Omega_p(W)$.

Finally, the defect of multiplicativity of Y is bounded by (4.17) uniformly in Z. Hence, by theorem 4.3, the p-variation of $\int \alpha(Z)\,dZ$ is controlled by ω times a constant which depends only on $\gamma, p, \omega(0,T)$ and $\|\alpha\|_{\mathrm{Lip}}$. $\qquad\square$

Finally, we are able to define the image of a rough path by a function.

Definition 4.13 (Image by a Function). *Let $Z : \triangle_T \longrightarrow T^{(\lfloor p \rfloor)}(V)$ be a p-rough path, which we assume to be geometric if $p \ge 3$. Let $f : V \longrightarrow W$ be a $\mathrm{Lip}(\gamma)$ function for some $\gamma > p$. Then $f(Z)$ is by definition the rough path $\int df(Z)\,dZ$ in $\Omega_p(W)$.*

5

Differential Equations Driven by Rough Paths

In this chapter, we finally make sense of a solution of a differential equation driven by a rough path and prove the existence and uniqueness of the solution under an assumption of smoothness on the vector field which is similar to that of Picard-Lindelöf's theorem (Theorem 1.28). As in the classical case, the proof relies on an iteration procedure which requires one more degree of smoothness on the vector field than what is needed to make sense of the equation. However, the iteration needed to handle rough paths is slightly more complicated than the usual one and we take some time to explain why. In particular, we point out some mistakes related to linear images and comparison of rough paths which we think one is likely to make when discovering the subject.

The proof presented here is neither exactly that of [60] nor that of [67] and we hope that it is somewhat simpler and easier to understand.

5.1 Linear Images of Geometric Rough Paths

There is a special case of the image of a geometric rough path by a regular function (see Definition 4.13) which we will use constantly in this chapter, namely the image by a linear mapping.

Let $A \in \mathbf{L}(V, W)$ be a continuous linear mapping between Banach spaces. Let $X : \triangle_T \longrightarrow T^{(\lfloor p \rfloor)}(V)$ be a geometric p-rough path in V. Then the rough path $A(X)$, defined by Definition 4.13, can be described in the following natural way. For each $k \geq 0$, A induces a linear mapping from $V^{\otimes k}$ to $W^{\otimes k}$, which sends $x_1 \otimes \ldots \otimes x_k$ to $Ax_1 \otimes \ldots \otimes Ax_k$. For each $n \geq 0$, by putting these linear mappings for $k = 0, \ldots, n$ together, we get a linear mapping from the truncated tensor algebra $T^{(n)}(V)$ to $T^{(n)}(W)$, denoted by $\Gamma(A)$. Then $A(X) : \triangle_T \longrightarrow T^{(\lfloor p \rfloor)}(W)$ is defined simply by the following relation:

$$A(X)_{s,t} = \Gamma(A)(X_{s,t}) \quad \forall (s,t) \in \triangle_T.$$

The case where A is a projection will be extremely useful to us and it deserves a warning. Consider a rough path Z in $G\Omega_p(V \oplus W)$. Let $\pi_V : V \oplus W \longrightarrow V$ and $\pi_W : V \oplus W \longrightarrow W$ denote the coordinate projections. Set $X = \pi_V(Z)$ and $Y = \pi_W(Z)$. Then X and Y are both rough paths, respectively, in V and W and we often write $Z = (X, Y)$. However, one must keep in mind that, whatever the notation suggests and in contrast with the case of classical paths, the rough path $Z = (X, Y)$ is not determined by X and Y. Indeed, the iterated integrals of Z contain cross-iterated integrals of X and Y which are not functions of X and Y separately. We will illustrate this point and some other counter-intuitive facts in Proposition 5.4 and the exercise thereafter.

Before that, we make sense of a solution of a differential equation driven by a rough path and state the main theorem of these notes.

5.2 Solution of a Differential Equation Driven by a Rough Path

Let, as usual, V and W denote Banach spaces, whose tensor powers are endowed with norms which satisfy the usual requirements of symmetry and consistency (see Definition 1.25). Let $\gamma > p \geq 1$ denote real numbers. Let $f : W \longrightarrow \mathbf{L}(V, W)$ be a Lip$(\gamma - 1)$ function. Consider $X \in G\Omega_p(V)$ and $\xi \in W$. When X has bounded variation, the equation

$$dY_t = f(Y_t) \, dX_t \,, \quad Y_0 = \xi \tag{5.1}$$

is equivalent, up to a translation of vector ξ in W, to the system of equations

$$\begin{cases} dY_t = f_\xi(Y_t) \, dX_t \,, & Y_0 = 0, \\ dX_t = & dX_t, \end{cases}$$

where $f_\xi(\cdot) = f(\cdot + \xi)$. Let us define the one-form $h : V \oplus W \longrightarrow \mathrm{End}(V \oplus W)$ by

$$h(x, y) = \begin{pmatrix} \mathrm{Id}_V & 0 \\ f_\xi(y) & 0 \end{pmatrix}. \tag{5.2}$$

Then, still when X has bounded variation, solving (5.1) is equivalent to finding Z in $\Omega_1(V \oplus W)$ such that

$$dZ_t = h(Z_t) \, dZ_t \,, \quad Z_0 = 0 \,, \text{ and } \pi_V(Z) = X.$$

The point of this reformulation is that it makes sense even when X is a genuine rough path.

Definition 5.1. *Let $f : W \longrightarrow \mathbf{L}(V, W)$ be a Lip$(\gamma - 1)$ function. Consider $X \in G\Omega_p(V)$ and $\xi \in W$. Set $f_\xi(\cdot) = f(\cdot + \xi)$. Define $h : V \oplus W \longrightarrow \mathrm{End}(V \oplus W)$ by (5.2). We call $Z \in G\Omega_p(V \oplus W)$ a solution of the differential equation*

$$dY_t = f(Y_t) \, dX_t \,, \quad Y_0 = \xi \tag{5.3}$$

if the following two conditions hold:

1. $Z = \int h(Z)\,dZ$,
2. $\pi_V(Z) = X$.

5.3 The Universal Limit Theorem

In order to find a solution to (5.3), we are going to use Picard iteration. Let $X \in G\Omega_p(V)$ and $\xi \in W$ be given. We define $Z(0) = (X, \mathbf{0}) \in G\Omega_p(V \oplus W)$, where $\mathbf{0}$ denotes the trivial rough path defined by $\mathbf{0}_{s,t} = (1, 0, 0, \dots)$. Of course, in this special case, the warning made at the end of Sect. 5.1 is irrelevant and $Z(0)$ is indeed determined by X. Then, we define, for all $n \geq 0$, $Z(n+1) = \int h(Z(n))\,dZ(n)$.

Lemma 5.2. *For all $n \geq 0$, $\pi_V(Z(n)) = X$.*

Proof. The result holds by definition for $n = 0$. Let us prove that, for all $Z \in G\Omega_p(V \oplus W)$, the following equality holds:

$$\pi_V(Z) = \pi_V\left(\int h(Z)\,dZ\right).$$

By the definition of h, this relation holds if Z belongs to $\Omega_1(V \oplus W)$. Since both terms are continuous functions of Z in the p-variation topology, the identity holds for all $Z \in G\Omega_p(V \oplus W)$. $\square$

Let us denote, for all $n \geq 0$, $Y(n) = \pi_W(Z(n))$. With the usual warning, we write this under the form $Z(n) = (X, Y(n))$.

Just as in the classical case, the proof that $(Y(n))_{n \geq 0}$ converges to the unique solution requires one more degree of smoothness on f than what is needed to make sense of the equation.

Theorem 5.3 (Universal Limit Theorem). *Let $p \geq 1$ and $\gamma > p$ be real numbers. Let $f : W \longrightarrow \mathbf{L}(V, W)$ be a Lip(γ) function. For all $X \in G\Omega_p(V)$ and all $\xi \in W$, the equation*

$$dY_t = f(Y_t)\,dX_t, \quad Y_0 = \xi \tag{5.4}$$

admits a unique solution $Z = (X, Y) \in G\Omega_p(V \oplus W)$ in the sense of Definition 5.1. This solution depends continuously on X and ξ and the mapping $I_f : G\Omega_p(V) \times W \longrightarrow G\Omega_p(W)$ which sends (X, ξ) to Y is the unique extension of the Itô map which is continuous in the p-variation topology.

The rough path Y is the limit of the sequence $(Y(n))_{n \geq 0}$ of rough paths defined earlier. Moreover, let ω be a control of the p-variation of X. For all

$\rho > 1$, *there exists* $T_\rho \in (0, T]$ *such that this convergence holds geometrically fast at rate* ρ *on* $[0, T_\rho]$. *More precisely,*

$$\|Y(n)_{s,t}^i - Y(n+1)_{s,t}^i\| \le 2^i \rho^{-n} \frac{\omega(s,t)^{\frac{i}{p}}}{\beta\left(\frac{i}{p}\right)!} \quad \forall (s,t) \in \triangle_{T_\rho}, \ \forall i = 0, \dots, \lfloor p \rfloor.$$

$$(5.5)$$

Finally, T_ρ *depends only on* $\|f\|_{\mathrm{Lip}(\gamma)}$, p, γ *and* ω.

5.4 Linear Images and Comparison of Rough Paths

In order to prove (5.5), which is the main part of the proof of Theorem 5.3, we need to compare $Y(n)$ and $Y(n+1)$ and it is natural to couple them to do this. It is indeed not hard to write $(X, Y(n), Y(n+1))$ as the integral of a one-form along $(X, Y(n-1), Y(n))$.

Before going further, let us point out once again that this formulation may be misleading. What we are claiming is exactly this: it is possible and not hard to find a one-form $h_1 : V \oplus W \oplus W \longrightarrow \mathrm{End}(V \oplus W \oplus W)$ and to define by recurrence a sequence of rough paths $(Z_1(n))_{n \ge 0}$ in $V \oplus W \oplus W$ by $Z_1(n+1) = \int h_1(Z_1(n)) \, dZ_1(n)$ such that for all $n \ge 0$, $\pi_V(Z_1(n)) = X$, $\pi_{W,1}(Z_1(n)) = Y(n)$ and $\pi_{W,2}(Z_1(n)) = Y(n+1)$, with the obvious definition for $\pi_{W,1}$ and $\pi_{W,2}$. However, there may be several distinct ways to do this. We will soon write down explicitly our choice of h_1.

For the moment, assume that $Y(n)$ and $Y(n+1)$ are coupled in a rough path $Z_1(n) = (X, Y(n), Y(n+1))$. In order to compare $Y(n)$ and $Y(n+1)$, it would be tempting to consider the rough path $Y(n) - Y(n+1)$, which is well defined as the image of $Z_1(n)$ by the linear mapping $(x, y_1, y_2) \mapsto y_1 - y_2$. However, this would be a mistake, as the following proposition shows.

Proposition 5.4 (Counter-Example). *Take* $\varepsilon > 0$. *There exists* $Z \in G\Omega_{2+\varepsilon}(\mathbb{R}^2 \oplus \mathbb{R}^2)$, *with* $Z = (X, Y)$, *such that* $X \ne Y$ *and* $X - Y = \mathbf{0}$.

Proof. The details of this proof are left as an exercise for the reader. For each $n \ge 1$, define $X(n)$ and $Y(n)$ in $\Omega_1(\mathbb{R}^2)$ by:

$$X(n)_t = \frac{1}{n}(\cos 2\pi n^2 t, \sin 2\pi n^2 t), \quad Y(n)_t = \frac{1}{n}(\cos 2\pi n^2 t, -\sin 2\pi n^2 t).$$

As n tends to infinity, the path $(X(n), Y(n)) \in \Omega_1(\mathbb{R}^2 \oplus \mathbb{R}^2)$ converges in $\Omega_{2+\varepsilon}(\mathbb{R}^2 \oplus \mathbb{R}^2)$ towards a rough path Z defined as follows. Let (e_1, e_2, e_3, e_4) be the canonical basis of $\mathbb{R}^2 \oplus \mathbb{R}^2$. We identify $(\mathbb{R}^2 \oplus \mathbb{R}^2)^{\otimes 2}$ with $M_4(\mathbb{R})$ by identifying $e_i \otimes e_j$ with the elementary matrix $E_{i,j}$. Then

$$Z_{s,t} = \left(1, 0, \frac{t-s}{2} \begin{pmatrix} 0 & 1 & 0 & -1 \\ -1 & 0 & -1 & 0 \\ 0 & 1 & 0 & -1 \\ 1 & 0 & 1 & 0 \end{pmatrix}\right) \in T^{(2)}(\mathbb{R}^2 \oplus \mathbb{R}^2).$$

In particular $X(n)$ and $Y(n)$ converge, respectively, to X and Y defined by:

$$X_{s,t} = \left(1, 0, \frac{t-s}{2} \begin{pmatrix} 0 & 1 \\ -1 & 0 \end{pmatrix}\right) \text{ and } Y_{s,t} = \left(1, 0, \frac{t-s}{2} \begin{pmatrix} 0 & -1 \\ 1 & 0 \end{pmatrix}\right).$$

This can be proved either by computing the limits of $X(n)$ and $Y(n)$ or by taking projections of Z. In any case, it appears that $X \neq Y$.

However, for each n, the path $X(n) - Y(n)$ in $\mathbb{R}^2$ stays on the y-axis. Hence, it encloses no area and the limit of $X(n) - Y(n)$ is the trivial rough path $(1, 0, 0)$. A direct computation shows also that the image of Z under the mapping $(x, y) \mapsto x - y$ is the trivial rough path. Hence, we have proved twice that $X - Y = \mathbf{0}$. $\qquad\square$

Exercise 5.5. Check that $\widetilde{Y}(n)$ defined by:

$$\widetilde{Y}(n)_t = \frac{1}{n}(\sin 2\pi n^2 t, \cos 2\pi n^2 t)$$

converges to the same rough path Y as $Y(n)$. However, the pair $(X(n), \widetilde{Y}(n))$ does not converge to Z, but rather to another coupling $\widetilde{Z}$ of X and Y defined by:

$$\widetilde{Z}_{s,t} = \left(1, 0, \frac{t-s}{2} \begin{pmatrix} 0 & 1 & 1 & 0 \\ -1 & 0 & 0 & -1 \\ -1 & 0 & 0 & -1 \\ 0 & 1 & 1 & 0 \end{pmatrix}\right).$$

Hence, $\widetilde{Z} = (X, Y)$ and $Z = (X, Y)$ but $\widetilde{Z} \neq Z$. In fact, $\widetilde{Z}$ is simply the image of Z by the linear mapping $(x, y) \mapsto (x, R_{\frac{\pi}{2}}(y))$, where $R_{\frac{\pi}{2}}$ is the rotation of angle $\frac{\pi}{2}$.

These examples indicate that $(X, Y(n), Y(n+1))$ does not contain enough information to allow us to compare easily $Y(n)$ and $Y(n + 1)$. The following result explains what extra information is needed and it should not come as a surprise that it is some cross-information between rough paths.

In the following lemma, E denotes a Banach space for which we have chosen tensor norms which fulfil the usual requirements (see Definition 1.25). Moreover, we put on $E \oplus E$ a norm such that the coordinate projections π_1 and π_2 from $E \oplus E$ to E have norm 1.

Lemma 5.6. *Consider* $Z = (X, Y) \in G\Omega_p(E \oplus E)$. *Fix* $\varepsilon > 0$. *Consider* $D : E \oplus E \longrightarrow E \oplus E$ *defined by* $D(x, y) = (x, \frac{y-x}{\varepsilon})$. *Let* $\omega : \triangle_T \longrightarrow [0, +\infty)$ *be a control. Assume that* $D(Z) = (X, \frac{Y-X}{\varepsilon})$ *is controlled by* ω. *Then*

$$\|X_{s,t}^i - Y_{s,t}^i\| \leq ((1+\varepsilon)^i - 1)\frac{\omega(s,t)^{\frac{i}{p}}}{\beta\left(\frac{i}{p}\right)!} \quad \forall(s,t) \in \triangle_T, \ \forall i = 0, \ldots, \lfloor p \rfloor.$$

Proof. Fix $(s,t) \in \triangle_T$ and $i \in \{0, \dots, \lfloor p \rfloor\}$. Assume that Z has bounded variation. Define H_0 and H_1 in $\Omega_1(E)$ by $H_0 = X$ and $H_1 = Y - X$. Then, the relation $Y_t = X_t + (Y - X)_t$ gives the following expression of $Y^i_{s,t}$:

$$Y^i_{s,t} = \sum_{k_1,\dots,k_i \in \{0,1\}} \int_{s<u_1<\dots<u_i<t} dH_{k_1,u_1} \dots dH_{k_i,u_i}$$

$$= X^i_{s,t} + \sum_{\substack{k_1,\dots,k_i \in \{0,1\} \\ k_1+\dots+k_i>0}} \varepsilon^{k_1+\dots+k_i} \pi_{k_1} \otimes \dots \otimes \pi_{k_i} D(Z)^i_{s,t}.$$

This last identity makes sense without the assumption that Z has bounded variation and remains true in this case by continuity. Hence, since $D(Z)$ is controlled by ω and the projections have norm 1, we have

$$\|X^i_{s,t} - Y^i_{s,t}\| \leq \sum_{\substack{k_1,\dots,k_i \in \{0,1\} \\ k_1+\dots+k_i>0}} \varepsilon^{k_1+\dots+k_i} \|D(Z)^i_{s,t}\|$$

$$\leq ((1+\varepsilon)^i - 1) \frac{\omega(s,t)^{\frac{i}{p}}}{\beta\left(\frac{i}{p}\right)!}$$

and this finishes the proof. $\qquad\qquad\qquad\qquad\qquad\qquad\qquad\qquad\qquad\qquad\square$

According to this lemma, it is sufficient to control $(Y(n), \rho^n(Y(n+1) - Y(n)))$ uniformly in n in order to prove the geometric convergence of the sequence $(Y(n))_{n \geq 0}$. We are going to prove that there is a uniform control for $(X, Y(n), Y(n+1), \rho^n(Y(n+1) - Y(n)))$ on a time interval which depends on ρ. Before we do this, let us explain how we couple all these rough paths.

5.5 Three Picard Iterations

Recall that $f_\xi : W \longrightarrow \mathbf{L}(V, W)$ is given and that it is $\mathrm{Lip}(\gamma)$ for some $\gamma > p$. According to the division property of Lip functions stated in Proposition 1.26, there exists $g : W \times W \longrightarrow \mathbf{L}(W, \mathbf{L}(V, W))$ which is $\mathrm{Lip}(\gamma - 1)$ and such that for all $y_1, y_2 \in W$,

$$f_\xi(y_2) - f_\xi(y_1) = g(y_1, y_2)(y_2 - y_1).$$

Let us choose once for all a real $\rho > 1$. Consider the following one-forms.

$$h_0 : V \oplus W \longrightarrow \mathrm{End}(V \oplus W)$$

$$h_0(x,y) = \begin{pmatrix} \mathrm{Id}_V & 0 \\ f_\xi(y) & 0 \end{pmatrix},$$

$$h_1 : V \oplus W \oplus W \longrightarrow \mathrm{End}(V \oplus W \oplus W)$$

$$h_1(x, y_1, y_2) = \begin{pmatrix} \mathrm{Id}_V & 0 & 0 \\ 0 & 0\,\mathrm{Id}_W \\ f_\xi(y_2) & 0 & 0 \end{pmatrix},$$

$$h_2 : V \oplus W \oplus W \oplus W \longrightarrow \mathrm{End}(V \oplus W \oplus W \oplus W)$$

$$h_2(x, y_1, y_2, d) = \begin{pmatrix} \mathrm{Id}_V & 0 & 0 & 0 \\ 0 & 0\,\mathrm{Id}_W & 0 \\ f_\xi(y_2) & 0 & 0 & 0 \\ \rho g(y_1, y_2)(d) & 0 & 0 & 0 \end{pmatrix}.$$

The form h_0 is just the same as the form h defined by (5.2). Recall that it allowed us to define $Z(n)$ by induction. We now denote this $Z(n)$ by $Z_0(n) = (X, Y(n))$. Let us explain what the forms h_1 and h_2 do.

Let us define $Z_1(n) \in \Omega_p(V \oplus W \oplus W)$ by:

$$Z_1(0) = (X, \mathbf{0}, Y(1)) \quad \text{and} \quad Z_1(n+1) = \int h_1(Z_1(n)) \, dZ_1(n).$$

Similarly, let us define $Z_2(n) \in \Omega_p(V \oplus W \oplus W \oplus W)$ by:

$$Z_2(0) = (X, \mathbf{0}, Y(1), Y(1)) \quad \text{and} \quad Z_2(n+1) = \int h_2(Z_2(n)) \, dZ_2(n).$$

Observe that h_2 is only $\mathrm{Lip}(\gamma - 1)$ and the definition of Z_2 uses the fact that $\gamma - 1 > p - 1$.

Lemma 5.7. *For all* $n \geq 0$, *the relations* $\pi_{W,1}(Z_1(n)) = Y(n)$ *and* $\pi_{W,2}(Z_1(n)) = Y(n+1)$ *hold. Similarly, for all* $n \geq 0$, *the relations* $\pi_{W,1}(Z_2(n)) = Y(n)$, $\pi_{W,2}(Z_2(n)) = Y(n+1)$ *and* $\pi_{W,3}(Z_2(n)) = \rho^n(Y(n+1) - Y(n))$ *hold.*

Here, by $Y(n+1) - Y(n)$ *we mean either the image of* $Z_1(n)$ *by the linear mapping* $(x, y_1, y_2) \mapsto y_2 - y_1$ *or the image of* $Z_2(n)$ *by the linear mapping* $(x, y_1, y_2, d) \mapsto y_2 - y_1$ *and this makes sense because these images agree.*

In other words, for all $n \geq 0$,

$$\begin{cases} Z_0(n) = (X, Y(n)), \\ Z_1(n) = (X, Y(n), Y(n+1)), \\ Z_2(n) = (X, Y(n), Y(n+1), \rho^n(Y(n+1) - Y(n))). \end{cases}$$

Proof. The proof of this lemma is really the same as that of Lemma 5.2. All the claimed equalities are true under the assumption that X has bounded variation. Since every term involved in these equalities depends continuously on X in the p-variation topology, the equalities hold without the extra smoothness assumption on X. ☐

The key point in the proof of Theorem 5.3 is the proof that the p-variation of $Z_2(n)$ is controlled uniformly with respect to n, on a time interval which depends on ρ. The main technical tool for proving this is a scaling lemma which we prove in the next paragraph.

5.6 The Main Scaling Result

Since we are going to use the scaling lemma several times for different spaces of rough paths, we present it in a neutral setting. So, let E and F be Banach spaces, whose tensor powers are endowed with norms which satisfy the usual requirements of symmetry and consistency (see Definition 1.25).

We are going to analyse rough paths in $E \oplus F$, so that tensor powers of $E \oplus F$ are going to appear. Let us discuss a very convenient choice of norm for these tensor powers.

First observe that, for all $n \geq 0$, the vector space $(E \oplus F)^{\otimes n}$ splits naturally into the direct sum of 2^n subspaces. To write this down, let us use the notation $G_0 = E$ and $G_1 = F$. Then

$$(E \oplus F)^{\otimes n} = \bigoplus_{\tau = (\tau_1, \ldots, \tau_n) \in \{0,1\}^n} G_{\tau_1} \otimes \ldots \otimes G_{\tau_n}.$$

Accordingly, every element of $(E \oplus F)^{\otimes n}$ can be written as $g = \sum_\tau g_\tau$. The norm of each vector g_τ is determined by the convention we choose on the norm on the tensor product of two Banach spaces, but we decide to endow $(E \oplus F)^{\otimes n}$ with the sup norm

$$\|g\| = \max_{\tau \in \{0,1\}^n} \|g_\tau\|.$$

As long as we work with finite-dimensional spaces, this does not affect the topologies in any way. In any case, this allows us a very convenient form of the following result.

Lemma 5.8 (Scaling Lemma). *Consider* $Z = (X, Y) \in G\Omega_p(E \oplus F)$. *Let* $\omega : \triangle_T \longrightarrow [0, +\infty)$ *be a control. Let* $M \geq 1$ *be a real number. Assume that the p-variation of X is controlled by ω and that of Z is controlled by $M\omega$.*

Then, for all ε such that $0 \leq \varepsilon \leq M^{-\frac{\lfloor p \rfloor}{p}}$, the p-variation of $(X, \varepsilon Y)$ is controlled by ω.

In this statement, $(X, \varepsilon Y)$ denotes as usual the image of Z by the linear mapping $(x, y) \mapsto (x, \varepsilon y)$.

Proof. Let us fix $\varepsilon \geq 0$. Let $S \in \mathrm{End}(E \oplus F)$ be the linear mapping defined by $S(x, y) = (x, \varepsilon y)$. Let us assume temporarily that Z has bounded variation. Then, setting $H_0 = X$ and $H_1 = Y$, the iterated integrals of $S(Z)$ can be written as follows. Choose $(s, t) \in \triangle_T$ and $i \in \{0, \ldots, \lfloor p \rfloor\}$. Let π_0 and π_1

denote, respectively, the projections from $E \oplus F$ onto the first and second factor. Then, with the notation $|\tau| = \tau_1 + \ldots + \tau_i$,

$$
\begin{aligned}
S(Z)^i_{s,t} &= \sum_{\tau=(\tau_1,\ldots,\tau_i)\in\{0,1\}^i} \varepsilon^{|\tau|} \int_{s<u_1<\ldots<u_i<t} dH_{\tau_1,u_1} \ldots dH_{\tau_i,u_i} \\
&= \sum_{\tau\in\{0,1\}^i} \varepsilon^{|\tau|} \pi_{\tau_1} \otimes \ldots \otimes \pi_{\tau_i} Z^i_{s,t} \\
&= X^i_{s,t} + \sum_{\tau\in\{0,1\}^i,|\tau|>0} \varepsilon^{|\tau|} \pi_{\tau_1} \otimes \ldots \otimes \pi_{\tau_i} Z^i_{s,t}.
\end{aligned}
$$

The familiar argument of continuity shows that the last equality holds without the assumption that Z has bounded variation, provided Z is geometric. Now, by our assumption on the norm on $(E \oplus F)^{\otimes i}$,

$$
\|S(Z)^i_{s,t}\| \leq \max\left(\|X^i_{s,t}\|, \max_{\tau\in\{0,1\}^i,|\tau|>0} \varepsilon^{|\tau|} \|Z^i_{s,t}\| \right).
$$

Since X is controlled by ω and Z by $M\omega$,

$$
\|S(Z)^i_{s,t}\| \leq \max\left(1, \max(\varepsilon, \varepsilon^{\lfloor p \rfloor}) M^{\frac{\lfloor p \rfloor}{p}}\right) \frac{\omega(s,t)^{\frac{i}{p}}}{\beta\left(\frac{i}{p}\right)!}.
$$

The result follows immediately. $\square$

From now on, we always assume that the tensor powers of direct sums of Banach spaces are endowed with sup norms.

5.7 Uniform Control of the Picard Iterations

We are now going to prove the uniform control result announced at the end of Sect. 5.5. It will then be easy to deduce a proof of Theorem 5.3.

Recall the definition of the one-forms h_0, h_1, h_2. Once $\rho > 1$ is fixed, these forms are well-defined. According to Theorem 4.12, there exists a constant M, which depends only on the Lip norms of h_0, h_1 and h_2, on p and on γ, such that the following holds: whenever Z is a rough path in the appropriate space with p-variation controlled by some control ω such that $\omega(0,T) \leq 1$, the p-variation of $\int h_i(Z) \, dZ$ is controlled by $M\omega$, where $i \in \{0,1,2\}$.

We may and do assume that $M \geq 1$. Then we set $\varepsilon = M^{-\frac{\lfloor p \rfloor}{p}}$. Now let ω_0 be a control of the p-variation of the signal X of the differential equation (5.4). Let $T_\rho > 0$ be such that $\omega_0(0,T_\rho) = \varepsilon^p$. Set $\omega = \varepsilon^{-p}\omega_0$. Then $\varepsilon^{-1}X$ is controlled by ω and $\omega(0,T_\rho) \leq 1$.

Proposition 5.9. *For all $n \geq 0$, the p-variation of the rough paths*

$$(\varepsilon^{-1}X, Y(n)), \tag{5.6}$$

$$(\varepsilon^{-1}X, Y(n), Y(n+1)), \tag{5.7}$$

$$and \ (\varepsilon^{-1}X, Y(n), Y(n+1), \rho^n(Y(n+1) - Y(n))) \tag{5.8}$$

is controlled by ω on $[0, T_\rho]$.

The rough paths (5.6), (5.7) and (5.8) are, respectively, defined as linear images of the rough paths $Z_0(n)$, $Z_1(n)$ and $Z_2(n)$. The bound on (5.6) must be proved first in order to prove others. The bound on (5.7) is not necessary for us, but we include it as a training for that on (5.8).

Proof. We prove this result successively for the three sequences of rough paths, each time by induction. To start with, the p-variation of $(\varepsilon^{-1}X, \mathbf{0})$ is controlled by ω on $[0, T_\rho]$ by assumption. Assume that this is true for $(\varepsilon^{-1}X, Y(n))$.

Let us consider the rough path (U_0, U_1) on $V \oplus W$ defined by:

$$(U_0, U_1) = \int h_0(\varepsilon^{-1}X, Y(n)) \, d(\varepsilon^{-1}X, Y(n)).$$

This integral can be computed easily by assuming first that X has bounded variation. We propose the following formal way of computing it: by definition of h_0,

$$\begin{cases} dU_0 = d(\varepsilon^{-1}X) = \varepsilon^{-1}dX \\ dU_1 = f_\xi(Y(n)) \, d(\varepsilon^{-1}X) = \varepsilon^{-1}f_\xi(Y(n)) \, dX. \end{cases}$$

Hence, (U_0, U_1) is nothing but $(\varepsilon^{-1}X, \varepsilon^{-1}Y(n+1))$.

Now, by definition of M, the p-variation of $(\varepsilon^{-1}X, \varepsilon^{-1}Y(n+1))$ is controlled by $M\omega$ on $[0, T_\rho]$. Since the p-variation of $\varepsilon^{-1}X$ is controlled by ω, we can use the scaling Lemma 5.8 to conclude that the p-variation of $(\varepsilon^{-1}X, Y(n+1))$ is controlled by ω on $[0, T_\rho]$. The first induction is completed.

The two other sequences are now treated exactly in the same way. Let us look at the second one, which is a linear image of $(Z_1(n))_{n \geq 0}$. The p-variation of $(\varepsilon^{-1}X, \mathbf{0}, Y(1))$ is controlled by ω on $[0, T_\rho]$ by the part of the proposition which we have already proved. Assume that the same is true for $(\varepsilon^{-1}X, Y(n), Y(n+1))$.

Let us consider the rough path (U_0, U_1, U_2) on $V \oplus W \oplus W$ defined by:

$$(U_0, U_1, U_2) = \int h_1(\varepsilon^{-1}X, Y(n), Y(n+1)) \, d(\varepsilon^{-1}X, Y(n), Y(n+1)).$$

Let us compute this integral with the same formal procedure as earlier:

$$\begin{cases} dU_0 = d(\varepsilon^{-1}X) = \varepsilon^{-1}dX \\ dU_1 = dY(n+1) \\ dU_2 = f_\xi(Y(n+1)) \, d(\varepsilon^{-1}X) = \varepsilon^{-1}f_\xi(Y(n+1)) \, dX. \end{cases}$$

Hence, $(U_0, U_1, U_2) = (\varepsilon^{-1}X, Y(n+1), \varepsilon^{-1}Y(n+2))$. Again, the validity of this computation can be checked by first assuming that X has bounded variation and then using the usual continuity argument.

Now, the p-variation of $(\varepsilon^{-1}X, Y(n+1), \varepsilon^{-1}Y(n+2))$ is controlled by $M\omega$ on $[0, T_\rho]$. Since, by the first part of the proposition, the p-variation of $(\varepsilon^{-1}X, Y(n+1))$ is controlled by ω, we can use the scaling Lemma 5.8 to conclude that the p-variation of $(\varepsilon^{-1}X, Y(n+1), Y(n+2))$ is controlled by ω on $[0, T_\rho]$. The second induction is completed.

Let us now consider the third sequence. The p-variation of the rough path $(\varepsilon^{-1}X, 0, Y(1), Y(1))$ is controlled by ω on $[0, T_\rho]$ by the first part of the proposition. Assume that the same holds for $(\varepsilon^{-1}X, Y(n), Y(n+1), \rho^n(Y(n+1) - Y(n)))$.

Let us define a rough path (U_0, U_1, U_2, U_3) on $V \oplus W \oplus W \oplus W$ by

$$(U_0, U_1, U_2, U_3) = \int h_2(\varepsilon^{-1}X, Y(n), Y(n+1), \rho^n(Y(n+1) - Y(n)))$$
$$\times d(\varepsilon^{-1}X, Y(n), Y(n+1), \rho^n(Y(n+1) - Y(n))).$$

The formal computation reads

$$\begin{cases} dU_0 = d(\varepsilon^{-1}X) = \varepsilon^{-1}dX \\ dU_1 = dY(n+1) \\ dU_2 = f_\xi(Y(n+1)) \, d(\varepsilon^{-1}X) = \varepsilon^{-1}f_\xi(Y(n+1)) \, dX \\ dU_3 = \rho g(Y(n), Y(n+1))(\rho^n(Y(n+1) - Y(n)))d(\varepsilon^{-1}X) \\ \qquad = \varepsilon^{-1}\rho^{n+1}(f_\xi(Y(n+1)) - f_\xi(Y_n))dX. \end{cases}$$

Hence,

$$(U_0, U_1, U_2, U_3) = (\varepsilon^{-1}X, Y(n+1), \varepsilon^{-1}Y(n+2), \varepsilon^{-1}\rho^{n+1}(Y(n+2) - Y(n+1))).$$

The p-variation of this rough path is controlled by $M\omega$ on $[0, T_\rho]$ and that of $(\varepsilon^{-1}X, Y(n+1))$ is controlled by ω on the same interval, by the first induction. Hence, the scaling lemma applies one more time and allows us to conclude that the p-variation of $(\varepsilon^{-1}X, Y(n+1), Y(n+2), \rho^{n+1}(Y(n+2) - Y(n+1)))$ is controlled by ω on $[0, T_\rho]$. This concludes the third and last induction. $\square$

Now that this strong result has been proved, it is not difficult to prove Theorem 5.3.

5.8 Proof of the Main Theorem

Existence – Since $\varepsilon^{-1} > 1$, it follows in particular from Proposition 5.9 that the p-variation of $Z_2(n) = (X, Y(n), Y(n+1), \rho^n(Y(n+1) - Y(n)))$ is controlled by ω on $[0, T_\rho]$ for all $n \geq 0$. Apply to $Z_2(n)$ the linear mapping

$(x, y_1, y_2, d) \mapsto ((x, y_1), (0, d))$ from $V \oplus W^{\oplus 3}$ to $(V \oplus W) \oplus (V \oplus W)$, which has norm 1. We find that, on $[0, T_\rho]$, the p-variation of

$$((X, Y(n)), \rho^n (0, Y(n+1) - Y(n))) = ((X, Y(n)), \rho^n [(X, Y(n+1)) - (X, Y(n))])$$

is controlled by ω. Hence, by Lemma 5.6 applied with $E = V \oplus W$, we have for all $(s, t) \in \triangle_{T_\rho}$ and all $i = 0, \ldots, \lfloor p \rfloor$ the inequality

$$\|(X, Y(n))^i_{s,t} - (X, Y(n+1))^i_{s,t}\| \le ((1 + \rho^{-n})^i - 1) \frac{\omega(s, t)^{\frac{i}{p}}}{\beta \left(\frac{i}{p} \right)!}$$

$$\le 2^i \rho^{-n} \frac{\omega(s, t)^{\frac{i}{p}}}{\beta \left(\frac{i}{p} \right)!}. \tag{5.9}$$

In particular, the sequence $((X, Y(n)))_{n \ge 0}$ converges in p-variation on $[0, T_\rho]$ to a rough path $(X, Y) \in G\Omega_p(V \oplus W)$, which is a solution of (5.4). This convergence holds at geometric speed and the inequality (5.5) follows from (5.9) or from Lemma 5.6 and the fact that $(Y(n), \rho^n (Y(n+1) - Y(n)))$ is controlled by ω on $[0, T_\rho]$.

In order to get a solution on the whole interval $[0, T]$, one needs to paste together local solutions. This is possible because, once ρ is fixed, T_ρ can be bounded below by a function of the usual parameters p, γ, $\|f\|_{\text{Lip}}$, and the modulus of continuity of ω on $[0, T]$. Finally, one gets the existence of a solution on the whole interval $[0, T]$.

Uniqueness – Assume that $\widetilde{Z} = (X, \widetilde{Y})$ is a solution of (5.4). In order to compare $Y(n)$ and $\widetilde{Y}$, we need to couple them. To this end, let us define $h_3 : V \oplus W \oplus W \oplus W \longrightarrow \text{End}(V \oplus W \oplus W \oplus W)$ by:

$$h_3(x, y, \tilde{y}, \tilde{d}) = \begin{pmatrix} \text{Id}_V & 0 & 0 & 0 \\ f_\xi(y) & 0 & 0 & 0 \\ 0 & 0 & \text{Id}_W & 0 \\ \rho g(y, \tilde{y})(\tilde{d}) & 0 & 0 & 0 \end{pmatrix}.$$

Set $Z_3(0) = (X, 0, \widetilde{Y}, \widetilde{Y})$ and, for all $n \ge 0$, $Z_3(n + 1) = \int h_3(Z_3(n)) \, dZ_3(n)$. Then $Z_3(n) = (X, Y(n), \widetilde{Y}, \rho^n(\widetilde{Y} - Y(n)))$. Now we can prove just as in Proposition 5.9 that the p-variation of $Z_3(n)$ is controlled uniformly in n on a small time interval. It follows then from Lemma 5.6 that $Y = \widetilde{Y}$ on this time interval. The uniqueness on $[0, T]$ follows easily, because the uniform continuity of ω allows us to bound from below the length of the time interval on which we are able to control $Z_3(n)$.

Continuity – Set $I_f(X, \xi) = (X, Y)$ and for each $n \ge 0$, $F_n(X, \xi) = (X, Y(n))$. We have proved that, for each $X \in G\Omega_p(V)$ and each $\xi \in W$, $F_n(X, \xi)$ converges to $I_f(X, \xi)$. Since the functions F_n are continuous on $G\Omega_p(V) \times W$, all we need to prove the continuity of I_f is to establish some uniformity in the convergence of F_n towards I_f.

Once again we need to see how T_ρ depends on X, f and ξ. We know that it depends on f only through the Lip-norms of h_0, h_1 and h_2. These norms depend only on the Lip-norms of f_ξ and g. Now $\|f_\xi\|_{\mathrm{Lip}} = \|f\|_{\mathrm{Lip}}$ for all ξ and $\|g\|_{\mathrm{Lip}} \le C\|f\|_{\mathrm{Lip}}$ for some constant C, according to Proposition 1.26. Hence, T_ρ depends in fact only on p, γ, $\|f\|_{\mathrm{Lip}}$ and ω.

If ω is a control, let $G\Omega_p^\omega(V)$ denote the set of geometric p-rough paths on V whose p-variation is controlled by ω. Then (5.9) shows that the convergence of F_n towards I_f is uniform on $G\Omega_p^\omega(V) \times W$, at least on a time interval which depends on ω. Again, the length of this interval can be bounded from below and the uniform convergence holds on $[0, T]$. Finally, if $(X(n))_{n \ge 0}$ is a sequence of rough paths which tends to X and if $\xi_n \to \xi$ then there exists a control ω such that all the $X(n)$ and X have p-variation controlled by ω. Hence, $I_f(X(n), \xi_n) \to I_f(X, \xi)$ and the mapping I_f is continuous. It is the unique continuous extension of the Itô map from $G\Omega_p(V) \times W$ to $G\Omega_p(V \oplus W)$.

References

Historic Papers

1. Kuo-Tsai Chen. Integration of paths – a faithful representation of paths by non-commutative formal power series. *Trans. Am. Math. Soc.*, 89:395–407, 1958.
2. Kuo-Tsai Chen. Iterated integrals and exponential homomorphisms. *Proc. Lond. Math. Soc.*, 4(3):502–512, 1954.
3. Kuo-Tsai Chen. Integration of paths, geometric invariants and a generalized Baker-Hausdorff formula. *Ann. Math.*, 65(2):163–178, 1957.
4. Kuo-Tsai Chen. Algebras of iterated path integrals and fundamental groups. *Trans. Am. Math. Soc.*, 156:359–379, 1971.
5. Michel Fliess. Fonctionnelles causales non linéaires et indéterminées non commutatives. *Bull. Soc. Math. France*, 109(1):3–40, 1981.
6. Wilhelm Magnus. *Algebraic aspects in the theory of systems of linear differential equations.* Research Rep. No. BR-3. Mathematics Research Group, Washington Square College of Arts and Science, New York University, 1953.
7. Wilhelm Magnus. On the exponential solution of differential equations for a linear operator. *Comm. Pure Appl. Math.*, 7:649–673, 1954.
8. Rimhak Ree. Lie elements and an algebra associated with shuffles. *Ann. Math.*, 68(2):210–220, 1958.
9. Laurence C. Young. An inequality of hölder type, connected with stieltjes integration. *Acta Math.*, 67:251–282, 1936.

Preprints, Papers in Progress and Submitted Papers

10. Denis Feyel and Arnaud de La Pradelle. Curvilinear integrals along rough paths. Preprint, 2003.
11. Peter Friz. Continuity of the Itô-map for Hölder rough paths with applications to the support theorem in Hölder norm. Preprint, 2003.
12. Peter Friz and Nicolas Victoir. A note on the notion of Geometric Rough Paths. Preprint, 2004.
13. Antoine Lejay. Stochastic differential equations driven by processes generated by divergence form operators. Preprint, 2002.

14. Antoine Lejay and Terry J. Lyons. On the importance of the Lévy area for studying the limits of functions of converging stochastic processes. Application to homogeneization. Preprint, 2003.
15. Antoine Lejay and Nicolas Victoir. On (p, q)-rough paths. Preprint, 2004.
16. Xiang D. Li and Terry J. Lyons. Smoothness of itô maps and simulated annealing on path spaces. Preprint, 2003.
17. Terry J. Lyons and Nicolas Victoir. An extension theorem to rough paths. Preprint, 2004.

Published Papers

18. Shigeki Aida. Semiclassical limit of the lowest eigenvalue of a Schrödinger operator on a Wiener space. *J. Funct. Anal.*, 203(2):401–424, 2003.
19. Shigeki Aida. Weak Poincaré inequalities on domains defined by Brownian rough paths. *Ann. Probab.*, 32(4):3116–3137, 2004.
20. Ole E. Barndorff-Nielsen and Neil Shephard. Realized power variation and stochastic volatility models. *Bernoulli*, 9(2):243–265, 2003.
21. Richard F. Bass, Ben Hambly, and Terry J. Lyons. Extending the Wong–Zakai theorem to reversible Markov processes. *J. Eur. Math. Soc.*, 4:237–269, 2002.
22. Fabrice Baudoin. Équations différentielles stochastiques conduites par des lacets dans les groupes de Carnot. *C. R. Math. Acad. Sci. Paris*, 338(9):719–722, 2004.
23. Fabrice Baudoin and Laure Coutin. Étude en temps petit des solutions d'EDS conduites par des mouvements browniens fractionnaires. *C. R. Math. Acad. Sci. Paris*, 341(1):39–42, 2005.
24. Hakima Bessaih, Massimiliano Gubinelli, and Francesco Russo. The evolution of a random vortex filament. *Ann. Probab.*, 33(5):1825–1855, 2005.
25. Zdzisław Brzeźniak and Remi Léandre. Horizontal lift of an infinite dimensional diffusion. *Potential Anal.*, 12(3):249–280, 2000.
26. Mireille Capitaine and Catherine Donati-Martin. The Lévy area process for the free Brownian motion. *J. Funct. Anal.*, 179(1):153–169, 2001.
27. Laure Coutin and Laurent Decreusefond. Abstract nonlinear filtering theory in the presence of fractional Brownian motion. *Ann. Appl. Probab.*, 9(4):1058–1090, 1999.
28. Laure Coutin and Antoine Lejay. Semi-martingales and rough paths theory. *Electron. J. Probab.*, 10(23),761–785 (electronic), 2005.
29. Laure Coutin and Zhongmin Qian. Stochastic analysis, rough path analysis and fractional Brownian motions. *Probab. Theory Relat. Field.*, 122(1):108–140, 2002.
30. Laurent Decreusefond. Stochastic integration with respect to Gaussian processses. *C. R. Math. Acad. Sci. Paris*, 334(10):903–908, 2002.
31. Mohammed Errami and Francesco Russo. n-covariation, generalized Dirichlet processes and calculus with respect to finite cubic variation processes. *Stoch. Process. Appl.*, 104(2):259–299, 2003.
32. Franco Flandoli, Massimiliano Gubinelli, Mariano Giaquinta, and Vincenzo M. Tortorelli. Stochastic currents. *Stoch. Process. Appl.*, 115(9):1583–1601, 2005.
33. Franco Flandoli and Francesco Russo. Generalized integration and stochastic ODEs. *Ann. Probab.*, 30(1):270–292, 2002.

34. David Freedman. *Brownian motion and diffusion*. Springer, Berlin Heidelberg New York, 2nd edition, 1983.
35. Peter Friz and Nicolas Victoir. Approximations of the Brownian rough path with applications to stochastic analysis. *Ann. Inst. H. Poincaré Probab. Stat.*, 41(4):703–724, 2005.
36. Jessica G. Gaines and Terry J. Lyons. Random generation of stochastic area integrals. *SIAM J. Appl. Math.*, 54(4):1132–1146, 1994.
37. Jessica G. Gaines and Terry J. Lyons. Variable step size control in the numerical solution of stochastic differential equations. *SIAM J. Appl. Math.*, 57(5):1455–1484, 1997.
38. Mihai Gradinaru, Francesco Russo, and Pierre Vallois. Generalized covariations, local time and Stratonovich Itô's formula for fractional Brownian motion with Hurst index $H \geq \frac{1}{4}$. *Ann. Probab.*, 31(4):1772–1820, 2003.
39. Mikhaïl Gromov. Carnot-Carathéodory spaces seen from within. In *Sub-Riemannian geometry*, volume 144 of *Progr. Math.*, pages 79–323. Birkhäuser, Basel, 1996.
40. Massimiliano Gubinelli. Controlling rough paths. *J. Funct. Anal.*, 216(1): 86–140, 2004.
41. Martin Hairer. Ergodicity of stochastic differential equations driven by fractional Brownian motion. *Ann. Probab.*, 33(2):703–758, 2005.
42. Ben Hambly and Terry J. Lyons. Stochastic area for Brownian motion on the Sierpinski gasket. *Ann. Probab.*, 26(1):132–148, 1998.
43. Ben Hambly and Terry J. Lyons. Uniqueness for the signature of a path of bounded variation and continuous analogues for the free group. Preprint, 2003.
44. Ben Hoff. *The Brownian frame process as a rough path*. Ph.D. Thesis, University of Oxford, 2005.
45. Yaozhong Hu. Integral transformations and anticipative calculus for fractional Brownian motions. *Mem. Am. Math. Soc.*, 175(825):viii+127, 2005.
46. Peter Imkeller and Christian Lederer. On the cohomology of flows of stochastic and random differential equations. *Probab. Theory Relat. Field.*, 120(2):209–235, 2001.
47. Peter Imkeller and Björn Schmalfuss. The conjugacy of stochastic and random differential equations and the existence of global attractors. *J. Dynam. Differential Equations*, 13(2):215–249, 2001.
48. F. Klingenhöfer and Martina Zähle. Ordinary differential equations with fractal noise. *Proc. Am. Math. Soc.*, 127(4):1021–1028, 1999.
49. Kestutis Kubilius. The existence and uniqueness of the solution of an integral equation driven by a p-semimartingale of special type. *Stoch. Process. Appl.*, 98(2):289–315, 2002.
50. Michel Ledoux, Zhongmin Qian, and Tusheng Zhang. Large deviations and support theorem for diffusion processes via rough paths. *Stoch. Process. Appl.*, 102(2):265–283, 2002.
51. Michel Ledoux, Terry Lyons, and Zhongmin Qian. Lévy area of Wiener processes in Banach spaces. *Ann. Probab.*, 30(2):546–578, 2002.
52. Antoine Lejay. On the convergence of stochastic integrals driven by processes converging on account of a homogenization property. *Electron. J. Probab.*, 7(18):18 (electronic), 2002.
53. Antoine Lejay. An introduction to rough paths. In *Séminaire de Probabilités XXXVII*, volume 1832 of *Lecture Notes in Math.*, pages 1–59. Springer, Berlin Heidelberg New York, 2003.

54. Jorge A. León and David Nualart. An extension of the divergence operator for Gaussian processes. *Stoch. Process. Appl.*, 115(3):481–492, 2005.
55. José R. León and Carenne Ludeña. Stable convergence of certain functionals of diffusions driven by fBm. *Stoch. Anal. Appl.*, 22(2):289–314, 2004.
56. Carenne Ludeña. Minimum contrast estimation for fractional diffusions. *Scand. J. Stat.*, 31(4):613–628, 2004.
57. Terry J. Lyons. On the nonexistence of path integrals. *Proc. R. Soc. Lond. A Math. Phys. Eng. Sci.*, 432(1885):281–290, 1991.
58. Terry J. Lyons. Differential equations driven by rough signals. I. An extension of an inequality of L.C. Young. *Math. Res. Lett.*, 1(4):451–464, 1994.
59. Terry J. Lyons. The interpretation and solution of ordinary differential equations driven by rough signals. In *Stochastic analysis (Ithaca, NY, 1993)*, volume 57 of *Proc. Symp. Pure Math.*, pages 115–128. Am. Math. Soc., Providence, RI, 1995.
60. Terry J. Lyons. Differential equations driven by rough signals. *Rev. Mat. Iberoamericana*, 14(2):215–310, 1998.
61. Terry J. Lyons and Zhongmin Qian. Calculus for multiplicative functionals, Itô's formula and differential equations. In *Itô's stochastic calculus and probability theory*, pages 233–250. Springer, Berlin Heidelberg New York Tokyo, 1996.
62. Terry J. Lyons and Zhongmin Qian. Calculus of variation for multiplicative functionals. In *New trends in stochastic analysis (Charingworth, 1994)*, pages 348–374. World Science, River Edge, NJ, 1997.
63. Terry J. Lyons and Zhongmin Qian. A class of vector fields on path spaces. *J. Funct. Anal.*, 145(1):205–223, 1997.
64. Terry J. Lyons and Zhongmin Qian. Flow equations on spaces of rough paths. *J. Funct. Anal.*, 149(1):135–159, 1997.
65. Terry J. Lyons and Zhongmin Qian. Stochastic Jacobi fields and vector fields induced by varying area on path spaces. *Probab. Theory Relat. Field.*, 109(4):539–570, 1997.
66. Terry J. Lyons and Zhongmin Qian. Flow of diffeomorphisms induced by a geometric multiplicative functional. *Probab. Theory Relat. Field.*, 112(1):91–119, 1998.
67. Terry J. Lyons and Zhongmin Qian. *System control and rough paths*. Oxford Mathematical Monographs. Oxford University Press, Oxford Science Publications, Oxford, 2002.
68. Terry J. Lyons and Nicolas Victoir. Stochastic analysis with applications to mathematical finance. Cubature on Wiener space. *Proc. R. Soc. Lond. A Math. Phys. Eng. Sci.*, 460(2041):169–198, 2004.
69. Terry J. Lyons and Ofer Zeitouni. Conditional exponential moments for iterated Wiener integrals. *Ann. Probab.*, 27(4):1738–1749, 1999.
70. Martynas Manstavičius. p-variation of strong Markov processes. *Ann. Probab.*, 32(3A):2053–2066, 2004.
71. Renaud Marty. Théorème limite pour une équation différentielle à coefficient aléatoire à mémoire longue. *C. R. Math. Acad. Sci. Paris*, 338(2):167–170, 2004.
72. Bohdan Maslowski and David Nualart. Evolution equations driven by a fractional Brownian motion. *J. Funct. Anal.*, 202(1):277–305, 2003.
73. Rimas Norvaiša. Chain rules and p-variation. *Stud. Math.*, 149(3):197–238, 2002.
74. Jiagang Ren and Xicheng Zhang. Path continuity of fractional Dirichlet functionals. *Bull. Sci. Math.*, 127(4):368–378, 2003.

75. Christophe Reutenauer. *Free Lie algebras*, volume 7 of *London Mathematical Society Monographs. New Series*. The Clarendon, Oxford Science Publications. Oxford University Press, New York, 1993.
76. Thomas Simon. Small ball estimates in p-variation for stable processes. *J. Theor. Probab.*, 17(4):979–1002, 2004.
77. Thomas Simon. Support theorem for jump processes. *Stoch. Process. Appl.*, 89(1):1–30, 2000.
78. Thomas Simon. Small deviations in p-variation for multidimensional Lévy processes. *J. Math. Kyoto Univ.*, 43(3):523–565, 2003.
79. E.-M. Sipiläinen. *A pathwise view of solutions of stochastic differential equations*. Ph.D. Thesis, University of Edinburgh, 1993.
80. Elias M. Stein. *Singular integrals and differentiability properties of functions*. Princeton Mathematical Series, No. 30. Princeton University Press, Princeton, NJ, 1970.
81. Constantin Tudor and Maria Tudor. On the two-parameter fractional Brownian motion and Stieltjes integrals for Hölder functions. *J. Math. Anal. Appl.*, 286(2):765–781, 2003.
82. Nicolas Victoir. Lévy area for the free Brownian motion: existence and non-existence. *J. Funct. Anal.*, 208(1):107–121, 2004.
83. David R. E. Williams. Path-wise solutions of stochastic differential equations driven by Lévy processes. *Rev. Math. Iberoamericana*, 17(2):295–329, 2001.
84. Martina Zähle. Integration with respect to fractal functions and stochastic calculus. II. *Math. Nachr.*, 225:145–183, 2001.

Index

List of Participants

Lecturers

CERF Raphaël	Univ. Paris-Sud, Orsay, F
LYONS Terry	Univ. Oxford, UK
SLADE Gordon	Univ. British Columbia, Vancouver, Canada

Participants

ASSELAH Amine	Univ. Provence, Marseille, F
AUTRET Solenn	Univ. Paul Sabatier, Toulouse, F
BAILLEUL Ismaël	Univ. Paris-Sud, Orsay, F
BAUDOIN Fabrice	Univ. Paul Sabatier, Toulouse, F
BEGYN Arnaud	Univ. Paul Sabatier, Toulouse, F
BEN-ARI Iddo	Technion Inst. Technology, Haifa, Israel
BERARD Jean	Univ. Lyon 1, F
BLACHE Fabrice	Univ. Blaise Pascal, Clermont-Ferrand, F
BROMAN Erik	Chalmers Univ. Techn., Gothenburg, Sweden
BROUTTELANDE Christophe	Univ. Paul Sabatier, Toulouse, F
BRYC Wlodzimierz	Univ. Cincinnati, USA
CARUANA Michael	Univ. Oxford, UK
CHIVORET Sebastien	Univ. Michigan, Ann Arbor, USA
COUPIER David	Univ. Paris 5, F
CROYDON David	Univ. Oxford, UK

DE CARVALHO BEZERRA S.   Univ. Henri Poincaré, Nancy, F

DE TILIERE Béatrice   Univ. Paris-Sud, Orsay, F

DELARUE François   Univ. Paris 7, F

DEVAUX Vincent   Univ. Rouen, F

DUNLOP François   Univ. Cergy-Pontoise, F

DUQUESNE Thomas   Univ. Paris-Sud, Orsay, F

FERAL Delphine   Univ. Paul Sabatier, Toulouse, F

FILLIGER Roger   EPFL, Lausanne, Switzerland

GARET Olivier   Univ. Orléans, F

GAUTIER Eric   Univ. Rennes 1 & INSEE, F

GOERGEN Laurent   ETH Zurich, Switzerland

GOUERE Jean-Baptiste   Univ. Claude Bernard, Lyon, F

GOURCY Mathieu   Univ. Blaise Pascal, Clermont-Ferrand, F

HOLROYD Alexander   Univ. British Columbia, Vancouver, Canada

ISHIKAWA Yasushi   Univ. Ehime, Matsuyama, Japan

JAKUBOWICZ Jérémie   ENS Cachan, F

JOULIN Aldéric   Univ. La Rochelle, F

KASPRZYK Arkadiusz   Univ. Wroclaw, Poland

KOVCHEGOV Yevgeniy   UCLA, Los Angeles, USA

KURT Noemi   Univ. Zurich, Switzerland

LACAUX Céline   Univ. Paul Sabatier, Toulouse, F

LACHAUD Béatrice   Univ. Paris 5, F

LE GALL Jean-Franccois   ENS Paris, F

LE JAN Yves   Univ. Paris-Sud, Orsay, F

LEI Liangzhen   Univ. Blaise Pascal, Clermont-Ferrand, F

LEVY Thierry   ENS Paris, F

LUCZAK Malwina   London School of Economics, UK

MARCHAND Régine   Univ. Henri Poincaré, Nancy, F

MARDIN Arif   TELECOM-INT, Evry, F

MARTIN James   Univ. Paris 7, F

MARTY Renaud   Univ. Paul Sabatier, Toulouse, F

MERLE Mathieu   ENS Paris, F

MESSIKH Reda Jürg	Univ. Paris-Sud, Orsay, F
MOCIOALCA Oana	Purdue Univ., West Lafayette, USA
NIEDERHAUSEN Meike	Purdue Univ., West Lafayette, USA
NINOMIYA Syoiti	Tokyo Instit. Technology, Japan
NUALART David	Univ. Barcelona, Spain
PICARD Jean	Univ. Blaise Pascal, Clermont-Ferrand, F
PUDLO Pierre	Univ. Claude Bernard, Lyon, F
RIVIERE Olivier	Univ. Paris 5, F
ROUSSET Mathias	Univ. Paul Sabatier, Toulouse, F
ROUX Daniel	Univ. Blaise Pascal, Clermont-Ferrand, F
SAINT LOUBERT BIE Erwan	Univ. Blaise Pascal, Clermont-Ferrand, F
SAVONA Catherine	Univ. Blaise Pascal, Clermont-Ferrand, F
SERLET Laurent	Univ. Paris 5, F
THOMANN Philipp	Univ. Zurich, Switzerland
TORRECILLA Iván	Univ. Barcelona, Spain
TRASHORRAS Jose	Univ. Warwick, Coventry, UK
TURNER Amanda	Univ. Cambridge, UK
TYKESSON Johan	Chalmers Univ. Techn., Göteborg, Sweden
VIGNAUD Yvon	CPT, Marseille, F
WEILL Mathilde	ENS Paris, F
WINKEL Matthias	Univ. Oxford, UK
YU Yuhua	Purdue Univ., West Lafayette, USA

List of Short Lectures

Delphine Feral	On large deviations for the spectral measure of both continuous and discrete Coulomb gas
Roger Filliger	Diffusion mediated transport
Olivier Garet	Asymptotic shape for the chemical distance and first-passage percolation on a Bernoulli cluster
Eric Gautier	Large deviations for stochastic nonlinear Schrödinger equations and applications
Jean-Baptiste Gouéré	Probabilistic characterization of quasicrystals
Alexander Holroyd	A stable marriage of Poisson and Lebesgue
Aldéric Joulin	Functional inequalities for continuous time Markov chains: a modified curvature criterion
Yevgeniy Kovchegov	Multi-particle systems with reinforcements
Céline Lacaux	Atypic examples of locally self-similar fields
Thierry Lévy	Large deviations for the Yang-Mills measure
Malwina Luczak	On the maximum queue length in the supermarket model
Régine Marchand	Coexistence in a simple competition model
James Martin	Heavy tails in last-passage percolation
Renaud Marty	Asymptotic behavior of a nonlinear Schrödinger equation in a random medium via a splitting method
Mathieu Merle	Local behaviour of the occupation time measure of super-Brownian motion
Reda Messikh	Large deviations in the supercritical vicinity of p_c
Oana Mocioalca	Skorohod integration and stochastic calculus beyond the fractional Brownian scale

David Nualart	Rough path analysis via fractional calculus
Pierre Pudlo	Precise estimates of large deviations of finite Markov chains with applications to biological sequence analysis
Mathias Rousset	Interacting particle systems approximations of Feynman-Kac formulae, applications to nonlinear filtering
Laurent Serlet	Super-Brownian motion conditioned to its total mass
José Trashorras	Fluctuations of the free energy in the high temperature Hopfield model
Yvon Vignaud	Magnetostriction
Matthias Winkel	Growth of the Brownian forest

Saint-Flour Probability Summer Schools

In order to facilitate research concerning previous schools we give here the names of the authors, the series*, and the number of the volume where their lectures can be found:

Summer School	Authors	Series	Vol. Nr.
1971	Bretagnolle; Chatterji; Meyer	LNM	307
1973	Meyer; Priouret; Spitzer	LNM	390
1974	Fernique; Conze; Gani	LNM	480
1975	Badrikian; Kingman; Kuelbs	LNM	539
1976	Hoffmann-Jörgensen; Liggett; Neveu	LNM	598
1977	Dacunha-Castelle; Heyer; Roynette	LNM	678
1978	Azencott; Guivarc'h; Gundy	LNM	774
1979	Bickel; El Karoui; Yor	LNM	876
1980	Bismut; Gross; Krickeberg	LNM	929
1981	Fernique; Millar; Stroock; Weber	LNM	976
1982	Dudley; Kunita; Ledrappier	LNM	1097
1983	Aldous; Ibragimov; Jacod	LNM	1117
1984	Carmona; Kesten; Walsh	LNM	1180
1985/86/87	Diaconis; Elworthy; Föllmer; Nelson; Papanicolaou; Varadhan	LNM	1362
1986	Barndorff-Nielsen	LNS	50
1988	Ancona; Geman; Ikeda	LNM	1427
1989	Burkholder; Pardoux; Sznitman	LNM	1464
1990	Freidlin; Le Gall	LNM	1527
1991	Dawson; Maisonneuve; Spencer	LNM	1541
1992	Bakry; Gill; Molchanov	LNM	1581
1993	Biane; Durrett;	LNM	1608
1994	Dobrushin; Groeneboom; Ledoux	LNM	1648
1995	Barlow; Nualart	LNM	1690
1996	Giné; Grimmett; Saloff-Coste	LNM	1665
1997	Bertoin; Martinelli; Peres	LNM	1717
1998	Emery; Nemirovski; Voiculescu	LNM	1738
1999	Bolthausen; Perkins; van der Vaart	LNM	1781
2000	Albeverio; Schachermayer; Talagrand	LNM	1816
2001	Tavaré; Zeitouni	LNM	1837
	Catoni	LNM	1851
2002	Tsirelson; Werner	LNM	1840
	Pitman	LNM	1875
2003	Dembo; Funaki	LNM	1869
	Massart	LNM	1896
2004	Cerf	LNM	1878
	Slade	LNM	1879
	Lyons**, Caruana, Lévy	LNM	1908
2005	Doney	LNM	1897
	Evans	Forthcoming	
	Villani	Forthcoming	

*Lecture Notes in Mathematics (LNM), Lecture Notes in Statistics (LNS)
**The St. Flour lecturer was T.J. Lyons.

Lecture Notes in Mathematics

For information about earlier volumes
please contact your bookseller or Springer
LNM Online archive: springerlink.com

Vol. 1815: A. M. Vershik (Ed.), Asymptotic Combinatorics with Applications to Mathematical Physics. St. Petersburg, Russia 2001 (2003)

Vol. 1816: S. Albeverio, W. Schachermayer, M. Talagrand, Lectures on Probability Theory and Statistics. Ecole d'Eté de Probabilités de Saint-Flour XXX-2000. Editor: P. Bernard (2003)

Vol. 1817: E. Koelink, W. Van Assche (Eds.), Orthogonal Polynomials and Special Functions. Leuven 2002 (2003)

Vol. 1818: M. Bildhauer, Convex Variational Problems with Linear, nearly Linear and/or Anisotropic Growth Conditions (2003)

Vol. 1819: D. Masser, Yu. V. Nesterenko, H. P. Schlickewei, W. M. Schmidt, M. Waldschmidt, Diophantine Approximation. Cetraro, Italy 2000. Editors: F. Amoroso, U. Zannier (2003)

Vol. 1820: F. Hiai, H. Kosaki, Means of Hilbert Space Operators (2003)

Vol. 1821: S. Teufel, Adiabatic Perturbation Theory in Quantum Dynamics (2003)

Vol. 1822: S.-N. Chow, R. Conti, R. Johnson, J. Mallet-Paret, R. Nussbaum, Dynamical Systems. Cetraro, Italy 2000. Editors: J. W. Macki, P. Zecca (2003)

Vol. 1823: A. M. Anile, W. Allegretto, C. Ringhofer, Mathematical Problems in Semiconductor Physics. Cetraro, Italy 1998. Editor: A. M. Anile (2003)

Vol. 1824: J. A. Navarro González, J. B. Sancho de Salas, $\mathscr{C}^\infty$ – Differentiable Spaces (2003)

Vol. 1825: J. H. Bramble, A. Cohen, W. Dahmen, Multiscale Problems and Methods in Numerical Simulations, Martina Franca, Italy 2001. Editor: C. Canuto (2003)

Vol. 1826: K. Dohmen, Improved Bonferroni Inequalities via Abstract Tubes. Inequalities and Identities of Inclusion-Exclusion Type. VIII, 113 p, 2003.

Vol. 1827: K. M. Pilgrim, Combinations of Complex Dynamical Systems. IX, 118 p, 2003.

Vol. 1828: D. J. Green, Gröbner Bases and the Computation of Group Cohomology. XII, 138 p, 2003.

Vol. 1829: E. Altman, B. Gaujal, A. Hordijk, Discrete-Event Control of Stochastic Networks: Multimodularity and Regularity. XIV, 313 p, 2003.

Vol. 1830: M. I. Gil', Operator Functions and Localization of Spectra. XIV, 256 p, 2003.

Vol. 1831: A. Connes, J. Cuntz, E. Guentner, N. Higson, J. E. Kaminker, Noncommutative Geometry, Martina Franca, Italy 2002. Editors: S. Doplicher, L. Longo (2004)

Vol. 1832: J. Azéma, M. Émery, M. Ledoux, M. Yor (Eds.), Séminaire de Probabilités XXXVII (2003)

Vol. 1833: D.-Q. Jiang, M. Qian, M.-P. Qian, Mathematical Theory of Nonequilibrium Steady States. On the Frontier of Probability and Dynamical Systems. IX, 280 p, 2004.

Vol. 1834: Yo. Yomdin, G. Comte, Tame Geometry with Application in Smooth Analysis. VIII, 186 p, 2004.

Vol. 1835: O.T. Izhboldin, B. Kahn, N.A. Karpenko, A. Vishik, Geometric Methods in the Algebraic Theory of Quadratic Forms. Summer School, Lens, 2000. Editor: J.-P. Tignol (2004)

Vol. 1836: C. Năstăsescu, F. Van Oystaeyen, Methods of Graded Rings. XIII, 304 p, 2004.

Vol. 1837: S. Tavaré, O. Zeitouni, Lectures on Probability Theory and Statistics. Ecole d'Eté de Probabilités de Saint-Flour XXXI-2001. Editor: J. Picard (2004)

Vol. 1838: A.J. Ganesh, N.W. O'Connell, D.J. Wischik, Big Queues. XII, 254 p, 2004.

Vol. 1839: R. Gohm, Noncommutative Stationary Processes. VIII, 170 p, 2004.

Vol. 1840: B. Tsirelson, W. Werner, Lectures on Probability Theory and Statistics. Ecole d'Eté de Probabilités de Saint-Flour XXXII-2002. Editor: J. Picard (2004)

Vol. 1841: W. Reichel, Uniqueness Theorems for Variational Problems by the Method of Transformation Groups (2004)

Vol. 1842: T. Johnsen, A. L. Knutsen, K_3 Projective Models in Scrolls (2004)

Vol. 1843: B. Jefferies, Spectral Properties of Noncommuting Operators (2004)

Vol. 1844: K.F. Siburg, The Principle of Least Action in Geometry and Dynamics (2004)

Vol. 1845: Min Ho Lee, Mixed Automorphic Forms, Torus Bundles, and Jacobi Forms (2004)

Vol. 1846: H. Ammari, H. Kang, Reconstruction of Small Inhomogeneities from Boundary Measurements (2004)

Vol. 1847: T.R. Bielecki, T. Björk, M. Jeanblanc, M. Rutkowski, J.A. Scheinkman, W. Xiong, Paris-Princeton Lectures on Mathematical Finance 2003 (2004)

Vol. 1848: M. Abate, J. E. Fornaess, X. Huang, J. P. Rosay, A. Tumanov, Real Methods in Complex and CR Geometry, Martina Franca, Italy 2002. Editors: D. Zaitsev, G. Zampieri (2004)

Vol. 1849: Martin L. Brown, Heegner Modules and Elliptic Curves (2004)

Vol. 1850: V. D. Milman, G. Schechtman (Eds.), Geometric Aspects of Functional Analysis. Israel Seminar 2002-2003 (2004)

Vol. 1851: O. Catoni, Statistical Learning Theory and Stochastic Optimization (2004)

Vol. 1852: A.S. Kechris, B.D. Miller, Topics in Orbit Equivalence (2004)

Vol. 1853: Ch. Favre, M. Jonsson, The Valuative Tree (2004)

Vol. 1854: O. Saeki, Topology of Singular Fibers of Differential Maps (2004)

Vol. 1855: G. Da Prato, P.C. Kunstmann, I. Lasiecka, A. Lunardi, R. Schnaubelt, L. Weis, Functional Analytic Methods for Evolution Equations. Editors: M. Iannelli, R. Nagel, S. Piazzera (2004)

Vol. 1856: K. Back, T.R. Bielecki, C. Hipp, S. Peng, W. Schachermayer, Stochastic Methods in Finance, Bressanone/Brixen, Italy, 2003. Editors: M. Fritelli, W. Runggaldier (2004)

Vol. 1857: M. Émery, M. Ledoux, M. Yor (Eds.), Séminaire de Probabilités XXXVIII (2005)

Vol. 1858: A.S. Cherny, H.-J. Engelbert, Singular Stochastic Differential Equations (2005)

Vol. 1859: E. Letellier, Fourier Transforms of Invariant Functions on Finite Reductive Lie Algebras (2005)

Vol. 1860: A. Borisyuk, G.B. Ermentrout, A. Friedman, D. Terman, Tutorials in Mathematical Biosciences I. Mathematical Neurosciences (2005)

Vol. 1861: G. Benettin, J. Henrard, S. Kuksin, Hamiltonian Dynamics – Theory and Applications, Cetraro, Italy, 1999. Editor: A. Giorgilli (2005)

Vol. 1862: B. Helffer, F. Nier, Hypoelliptic Estimates and Spectral Theory for Fokker-Planck Operators and Witten Laplacians (2005)

Vol. 1863: H. Führ, Abstract Harmonic Analysis of Continuous Wavelet Transforms (2005)

Vol. 1864: K. Efstathiou, Metamorphoses of Hamiltonian Systems with Symmetries (2005)

Vol. 1865: D. Applebaum, B.V. R. Bhat, J. Kustermans, J. M. Lindsay, Quantum Independent Increment Processes I. From Classical Probability to Quantum Stochastic Calculus. Editors: M. Schürmann, U. Franz (2005)

Vol. 1866: O.E. Barndorff-Nielsen, U. Franz, R. Gohm, B. Kümmerer, S. Thorbjønsen, Quantum Independent Increment Processes II. Structure of Quantum Lévy Processes, Classical Probability, and Physics. Editors: M. Schürmann, U. Franz, (2005)

Vol. 1867: J. Sneyd (Ed.), Tutorials in Mathematical Biosciences II. Mathematical Modeling of Calcium Dynamics and Signal Transduction. (2005)

Vol. 1868: J. Jorgenson, S. Lang, $Pos_n(R)$ and Eisenstein Series. (2005)

Vol. 1869: A. Dembo, T. Funaki, Lectures on Probability Theory and Statistics. Ecole d'Eté de Probabilités de Saint-Flour XXXIII-2003. Editor: J. Picard (2005)

Vol. 1870: V.I. Gurariy, W. Lusky, Geometry of Müntz Spaces and Related Questions. (2005)

Vol. 1871: P. Constantin, G. Gallavotti, A.V. Kazhikhov, Y. Meyer, S. Ukai, Mathematical Foundation of Turbulent Viscous Flows, Martina Franca, Italy, 2003. Editors: M. Cannone, T. Miyakawa (2006)

Vol. 1872: A. Friedman (Ed.), Tutorials in Mathematical Biosciences III. Cell Cycle, Proliferation, and Cancer (2006)

Vol. 1873: R. Mansuy, M. Yor, Random Times and Enlargements of Filtrations in a Brownian Setting (2006)

Vol. 1874: M. Yor, M. Émery (Eds.), In Memoriam Paul-André Meyer - Séminaire de probabilités XXXIX (2006)

Vol. 1875: J. Pitman, Combinatorial Stochastic Processes. Ecole d'Eté de Probabilités de Saint-Flour XXXII-2002. Editor: J. Picard (2006)

Vol. 1876: H. Herrlich, Axiom of Choice (2006)

Vol. 1877: J. Steuding, Value Distributions of L-Functions (2007)

Vol. 1878: R. Cerf, The Wulff Crystal in Ising and Percolation Models, Ecole d'Eté de Probabilités de Saint-Flour XXXIV-2004. Editor: Jean Picard (2006)

Vol. 1879: G. Slade, The Lace Expansion and its Applications, Ecole d'Eté de Probabilités de Saint-Flour XXXIV-2004. Editor: Jean Picard (2006)

Vol. 1880: S. Attal, A. Joye, C.-A. Pillet, Open Quantum Systems I, The Hamiltonian Approach (2006)

Vol. 1881: S. Attal, A. Joye, C.-A. Pillet, Open Quantum Systems II, The Markovian Approach (2006)

Vol. 1882: S. Attal, A. Joye, C.-A. Pillet, Open Quantum Systems III, Recent Developments (2006)

Vol. 1883: W. Van Assche, F. Marcellàn (Eds.), Orthogonal Polynomials and Special Functions, Computation and Application (2006)

Vol. 1884: N. Hayashi, E.I. Kaikina, P.I. Naumkin, I.A. Shishmarev, Asymptotics for Dissipative Nonlinear Equations (2006)

Vol. 1885: A. Telcs, The Art of Random Walks (2006)

Vol. 1886: S. Takamura, Splitting Deformations of Degenerations of Complex Curves (2006)

Vol. 1887: K. Habermann, L. Habermann, Introduction to Symplectic Dirac Operators (2006)

Vol. 1888: J. van der Hoeven, Transseries and Real Differential Algebra (2006)

Vol. 1889: G. Osipenko, Dynamical Systems, Graphs, and Algorithms (2006)

Vol. 1890: M. Bunge, J. Funk, Singular Coverings of Toposes (2006)

Vol. 1891: J.B. Friedlander, D.R. Heath-Brown, H. Iwaniec, J. Kaczorowski, Analytic Number Theory, Cetraro, Italy, 2002. Editors: A. Perelli, C. Viola (2006)

Vol. 1892: A. Baddeley, I. Bárány, R. Schneider, W. Weil, Stochastic Geometry, Martina Franca, Italy, 2004. Editor: W. Weil (2007)

Vol. 1893: H. Hanßmann, Local and Semi-Local Bifurcations in Hamiltonian Dynamical Systems, Results and Examples (2007)

Vol. 1894: C.W. Groetsch, Stable Approximate Evaluation of Unbounded Operators (2007)

Vol. 1895: L. Molnár, Selected Preserver Problems on Algebraic Structures of Linear Operators and on Function Spaces (2007)

Vol. 1896: P. Massart, Concentration Inequalities and Model Selection, Ecole d'Eté de Probabilités de Saint-Flour XXXIII-2003. Editor: J. Picard (2007)

Vol. 1897: R. Doney, Fluctuation Theory for Lévy Processes, Ecole d'Eté de Probabilités de Saint-Flour XXXV-2005. Editor: J. Picard (2007)

Vol. 1898: H.R. Beyer, Beyond Partial Differential Equations, On linear and Quasi-Linear Abstract Hyperbolic Evolution Equations (2007)

Vol. 1899: Séminaire de Probabilités XL. Editors: C. Donati-Martin, M. Émery, A. Rouault, C. Stricker (2007)

Vol. 1900: E. Bolthausen, A. Bovier (Eds.), Spin Glasses (2007)

Vol. 1901: O. Wittenberg, Intersections de deux quadriques et pinceaux de courbes de genre 1, Intersections of Two Quadrics and Pencils of Curves of Genus 1 (2007)

Vol. 1902: A. Isaev, Lectures on the Automorphism Groups of Kobayashi-Hyperbolic Manifolds (2007)

Vol. 1903: G. Kresin, V. Maz'ya, Sharp Real-Part Theorems (2007)

Vol. 1904: P. Giesl, Construction of Global Lyapunov Functions Using Radial Basis Functions (2007)

Vol. 1905: C. Prévôt, M. Röckner, A Concise Course on Stochastic Partial Differential Equations (2007)

Vol. 1906: T. Schuster, The Method of Approximate Inverse: Theory and Applications (2007)

Vol. 1907: M. Rasmussen, Attractivity and Bifurcation for Nonautonomous Dynamical Systems (2007)

Vol. 1908: T.J. Lyons, M. Caruana, T. Lévy, Differential Equations Driven by Rough Paths, Ecole d'Eté de Probabilités de Saint-Flour XXXIV-2004. (2007)

Recent Reprints and New Editions

Vol. 1618: G. Pisier, Similarity Problems and Completely Bounded Maps. 1995 – 2nd exp. edition (2001)

Vol. 1629: J.D. Moore, Lectures on Seiberg-Witten Invariants. 1997 – 2nd edition (2001)

Vol. 1638: P. Vanhaecke, Integrable Systems in the realm of Algebraic Geometry. 1996 – 2nd edition (2001)

Vol. 1702: J. Ma, J. Yong, Forward-Backward Stochastic Differential Equations and their Applications. 1999 – Corr. 3rd printing (2007)

Vol. 830: J.A. Green, Polynomial Representations of GL_n, with an Appendix on Schensted Correspondence and Littelmann Paths by K. Erdmann, J.A. Green and M. Schocker 1980 – 2nd corr. and augmented edition (2007)